Responses from readers, full text.

"Shaun Johnston raises many fascinating questions, asking how it is that many scientists seem to find it so embarrassingly difficult to think about any sort of consciousness, especially their own, that they'd rather say it isn't there? This touching coyness over such intimate matters surely does credit to their modesty, but it really doesn't help the rest of us as we struggle to understand human life.

"Of late, some of them have indeed begun to admit cautiously that consciousness is a Problem, which chiefly means that they no longer know quite where to hide it. But they badly need to be pressed to roll up their sleeves and face it directly. So, all good wishes to Johnston as he pesters them to get over their scruples!"

> **Mary Midgley**, Gifford Lecture 1989-90 *Science As Salvation* and author of *Evolution as a Religion: Strange Hopes and Stranger Fears.*

"The ranks of those biologists who oppose the concept that natural selection has a causal role in evolution have been growing recently. 'Hard' selectionists have reformed themselves as 'soft' selectionists. But this is not enough. While no one would argue that selection as 'differential survival and reproduction' does not exist, there are examples of evolutionary phenomena where it does not, and no logical basis for treating it as any kind of evolutionary cause.

"In *Save Our Selves from Science Gone Wrong* Shaun Johnston presents selectionism as the science that has gone wrong, corrupting biology as well as pertinent aspects of sociology, philosophy, and the popular media. Such a strong voice deserves attention.

"Responding to the question 'What do you put in its place?' Johnston begins at the top, with the consciousness of the individual self as the driver of evolution. This raises other questions about how organisms/selves with a lower capacity for consciousness were able to drive their own evolution in an organized manner. Johnston's creative response is to imbue the genome with selfhood and to drive evolution from the bottom up, as well as from the top down—a matter of 'self-design' rather than creative design."

> **Robert G. B. Reid**, Emeritus Professor of Biology at the University of Victoria British Columbia and author of *Evolutionary Theory: The Unfinished Synthesis* and *Biological Emergences: Evolution by Natural Experiment.*

"Your take on Darwinism in *Save Our Selves*, like all your writings, is lively and provocative. Good luck with it."

> **John Horgan**, author of *The End of Science: Facing the Limits of Knowledge in the Twilight of the Scientific Age.*

ALSO BY SHAUN JOHNSTON

Me and The Genies (novel)
Background to the creationism-natural selection controversy

Father, In a Far Distant Time I Find You (phantasy)
4000 years of evolutionary theory reshaping human nature

Save Our Selves

from science gone wrong

PHYSICALISM AND NATURAL SELECTION

Afterword:

Autobiography of an evolved self

by
Shaun Johnston

EVOLVED SELF PUBLISHING

Save Our Selves From Science Gone Wrong:
Physicalism and Natural Selection
by Shaun Johnston

Published by
Evolved Self Publishing
www.evolvedself.com

Evolved Self Publishing is an imprint of
Shaun Johnston Design Limited

ISBN: 0-9779470-2-5

Contents

Preface

This book is a quick easy-to-read manifesto. It tells you why natural selection's dangerous and how to challenge it. No, this book isn't about creationism. Instead it calls for a new "self-based" science that puts conscious decision-making back into human nature.

To me, that's what's most important, our conscious self. So I'm upset when I hear scientists say, "Conscious decision-making is just an illusion. Your consciousness can't actually make anything happen in the real world."

Where do ideas like that come from? They come from what made science go wrong in the first place. The "discovery" of natural selection just made everything worse. If consciousness can't make anything happen in the environment, then it can't be selected for, therefore it can't evolve, therefore it can't exist! Wherever natural selection goes, this implicit denial of consciousness goes along with it.

In the coming pages I'll show you where science went wrong and why natural selection can't be the mechanism driving evolution. We'll look at the assumptions behind it that get in the way of exploring alternatives. Then I'll show you how easy it is to come up with mechanisms for evolution that work better and don't deny consciousness.

A better theory of evolution—could that make any difference? It could make a huge difference. Imagine knowing the real mechanism behind evolution's awesome creative powers and being able to put those powers to work in your own self. You might discover possibilities inside you beyond your wildest dreams. Imagine science discovering how to really explore the self and the amazing breakthroughs that could lead to.

You won't need any special knowledge to follow along. You'll find more information about some topics in an appendix at the back of the book, along with recommended readings, notes and sources for the text, and an index.

Part 1

How present-day science threatens the self

CHAPTER 1

The self in danger

Your "self"—isn't that the last thing you'd think anyone could take from you? Yet I'm assuring you it's in great danger. Someone's planning to take it from you. And they'll take it from you without warning, without you even suspecting. They don't stand out—they're not Godzillas or little green men from Mars you can identify and run away from. They're mild unassuming folk you'd pass in the street without a second thought. Scientists.

The problem doesn't start with scientists. I trained as a scientist myself, and I can't imagine a more worthwhile career. The threat comes not from any fault of scientists but from science itself, from the mission science inherited from its foundations and the tools it built to carry out that mission. Together, that mission and those tools created a comprehensive understanding of the universe that—even though it denies the self—to insiders seems complete. The result, as I'll show you, is a science that tells us the self doesn't exist.

It's a devastating denial. It threatens to accompany science into every culture in the world and imperil all existing traditions of human nature. Should you care? Only if having a conscious self is important to you. It is? Then read on.

What you'll find in these pages is a call to defend conscious experience. Sometimes I'll be talking about history—the history of science for example—other times I'll be talking about science itself. But always I'll be pointing out how today's science threatens the self and our ability to come to our own decisions in consciousness.

The theory behind the threat: "Physicalism"

Where does science find spokespeople to deny the conscious self? According to one authority on consciousness, maybe as many as a third of university students and teachers don't see consciousness as anything that needs accounting for. To them, this "consciousness" they hear other people talking about sounds like some kind of mystical experience. Often—too often—they write long books explaining how consciousness is really an illusion. Here's Daniel Dennett, a well-known writer on the subject: "Our minds are just what our brains non-miraculously do.... We are each made of mindless robots and nothing else, no non-physical, non-robotic ingredients at all." He's written several books trying to prove to the rest of us that we're mistaken and if we'd only give our eyes a good good rub the world would end up looking the same to us as it does to him.

Science's denial of consciousness has had many names: first Materialism and Reductionism then Emergence and today "Physicalism." Physicalism doesn't say the conscious self doesn't exist, all it makes claims about is the cause of things. In the physical world, it claims, only physical processes can make anything happen. Because consciousness isn't physical it can't work through physical processes, so it can't change anything in the real world. It can't, for example, start a war!

Accept that, and you've a stark choice. You can say, conscious decision-making isn't physical but then it can't make anything happen in the "real" world. Or you can say conscious decision-making is physical, it's just brain chemistry. Then it can make things happen in the physical world but it's just chemistry and isn't "free" any more than any other kind of chemistry is "free." You've no free will.

That second choice, that conscious decision-making is just something happening in brain chemistry, is what science tells us. Brain chemistry is where your decisions really get made. So your conscious self isn't free to make its "own" decisions. It can't really be held responsible for the decisions you make. And this is not only discussed by academics in universities, it's starting to infiltrate the school classroom.

I beg to differ. I believe the conscious self is not just brain chemistry, it's not just mindless robots. At the same time I believe it can make a difference in the world.

A little science about the self

Let's do some experiments on the self. Turn over this book and look at the back cover. Did you do that? You decided either to do it, or not to do it. Did you decide consciously? Or was the decision made for you in brain chemistry?

Here's how to tell. Didn't you reach your decision through a series of thoughts? Weren't those thoughts connected to one another by their meanings? They weren't connected by chemical reactions, at least not chemical reactions as most scientists think of them today.

Here's another experiment. If your self, your conscious self, can't drive physical processes, then it can't make anything happen in the brain, and the brain can't know anything about it. The brain can't know what conscious decision-making feels like. Now think back to your decision to look at this book's back cover? Did you experience making that decision consciously? If so, move your foot. If you moved your foot, you made something happen in the real world because of something you experienced in consciousness. If you're too lazy to move your foot, just say, "yes." Or just think "yes."

Did you think, "yes"? Then your consciousness can drive brain chemistry.

That's how easy it is to do research on consciousness. We come up with a question, think about it a bit, then come to a conscious decision. One thing this tells us is, doing science involves conscious decision-making. If someone tells you something is science, don't you assume someone somewhere made a conscious judgment about it, that it wasn't just spun off from a series of chemical reactions? When scientists say, "There's no such thing as conscious decision-making, it's really an illusion," you can reply, "That's OK. All I need is whatever you scientists use to decide what's true in science and what isn't. I'll settle for whatever that is." They can't have it both ways. Either they use conscious judgment to arrive at what's true in science, or they don't. And if they don't, if they're just robots, then what's the point of science? Why bother with what it says?

My mission—to rescue consciousness

If you're wondering when the arguments for creationism are going to appear, you won't get much out of this book. I don't believe in creationism. I've no interest in it. That's not my mission.

My mission is to save our selves from the wrong kind of science.

What matters to me most is the actual experience of consciousness and how I can change that experience "at will." For example, sometimes when I catch myself rushing somewhere I'll turn round and look back the way I've come. I do that just for the conscious experience of a less focused attention, so I can enjoy directing my attention anywhere I want, without focusing on any particular destination. Being able to direct my attention like that is just one example of why I value being able to "consciously" alter my conscious experience.

Though what I care about most is conscious experience what you'll find me talking about is "conscious decision-making." If consciousness can't drive brain chemistry to make things happen in the outside world, I can't claim it can drive brain chemistry to create its own experiences inside. Of course, our brain chemistry does sometimes drive our consciousness, but our consciousness can turn right around and drive our brain chemistry.

Does all this talk about consciousness make any difference? Sure does. It's the difference between saying:

"It doesn't matter what I think, everything's going to turn out just the same, thinking about things isn't going to change them."

and

"By making conscious decisions I can make things happen in the world that otherwise wouldn't happen, so consciously thinking about things is a huge responsibility."

It literally makes all the difference in the world. Not only in your own world, but everywhere in the world that's affected by your decisions.

Does this help you figure out where you stand? If like me you value having a conscious self, free to direct its own attention, free to make its own decisions, turn to Chapter 2 for where science went wrong. To learn more about Physicalism and brain chemistry check out the "Fine Print" Appendix starting on page 94.

CHAPTER 2

How science came to deny the self

What a good idea science is! Gather information direct from nature. Come up with different theories to account for it. Do experiments to see which theory is right. Write your conclusions up so other people can verify them. Gather those conclusions together into broader and broader generalizations. The whole process seems thoroughly rational. So it's surprising that modern science got its start in one religion, and then got taken over by another. You may not have been told about this in science class but it has a lot to do with where science went wrong.

400 years ago Europe was in turmoil. The unthinkable had happened—the Christian religion had split into two, Catholic and Protestant. Wild new forms of religion were springing up. Ancient Roman magic was being revived. Weaving together mystical practices from all over the Ancient world, "magicians" devised talismans resembling Tarot cards to divine the future. Alchemists chanted spells to draw heavenly influence down into their alembics. From many directions scholars were converging on a great quest—to create science.

They knew what they were after—the ultimate source of power. They'd seen examples of it in the introduction of clockworks, gunpowder, the printing press, the magnetic compass. In the Middle Ages, what were called the "mechanical arts" were practiced as a form of Christian worship—think of the immense skill involved in the building and embellishing of immense cathedrals with their flying buttresses, elaborate stone carvings, huge stained glass windows, and masterworks in silver and innumerable other of those "arts." At Clairvaux, the principal abbey of the Cistercians, canals carried water around the abbey to water-powered "factories" for milling, tanning and black-smithing, the water "seeking every task" as one visitor put it. So people had plenty of examples of this new power around them. The question was, how could you, whenever you wanted to, come up with new ones?

What they were looking for was the modern industrial research lab. And in 1662 in England it got invented. It was called The Royal Society. And it laid the foundations for England's lead in the "mechanical arts" of mining, ship building, navigation, ballistics and everything else you'd need to know to found a mighty empire, greater than the world had ever seen which, two centuries later, it brought into being.

So our question becomes, what led to this breakthrough in England in 1662? The answer is surprising. It begins centuries before. And it involves religion.

Grandchild of Millennialism

We start way back in the Middle Ages, at the time of the Crusades, with a Cistercian abbot called Joachim. Joachim was studying the Bible's Book of Revelation when he was interrupted by a vision. In this vision God revealed to Joachim His plan for the World, which not surprisingly involved what Joachim had just been reading about, the end of the Millennium and the coming of the "Apocalypse." On the strength of this vision Joachim founded a new "order" and began training a saintly vanguard that, serving as the "elect," would precipitate the Apocalypse.

Never heard of this Joachim? Me neither. Yet according to someone who has, Joachim's message "ignited the greatest spiritual revolution of the Middle Ages." It was the "most influential prophetic system known to Europe until Marxism." That covers about six and a half centuries, with the origin of science coming about two-thirds of the way through.

Joachim's message seems a bit wooly today, and you may wonder what it could possibly have to do with science. But it will mysteriously recur when we least expect it.

Joachim divided history leading up to the Millennium into three stages, one for each person of the Trinity. First stage: Father, representing marriage and the family and beginning with Adam. Second stage: Son, beginning with Christ, representing the church and its priests. Third stage: the Holy Ghost, representing the monasteries. According to Joachim the appearance of monks forming a saintly vanguard signaled the coming of the Millennium. With the coming of the Millennium, through their spiritual contemplation and teaching this saintly vanguard would redeem all mankind.

With Joachim, Christianity actually split into two religions. The original one proclaimed by St Paul was about humans qualifying to get to Heaven one by one. This new religion, Millennialism, was about the entire human race getting saved in one go through the efforts of a spiritual elite.

And spiritual elites promptly did appear, such as the Franciscans, one of whom—greatly influenced by Joachim—was the legendary English scholar Roger Bacon.

Here's what one authority says about him:

If he recognized the practical potential of natural philosophy [science], urged greater development of the [mechani-
cal] arts, and envisioned such modern inventions as self-
powered cars, boats, submarines, and airplanes, he did so
only with reference to the end-times, which he believed
were already at hand.

Already here you can see the ingredients of science coming
together, but for the sake of saving all mankind.

According to the Old Testament, mankind was original-
ly made in God's image. At first, Adam shared in God's wisdom. But with sin and then banishment from the Garden,
Adam and all his descendents lost access to that wisdom
and our perfect God-like image got blemished. If human-
kind could recover that lost wisdom, the story went, our
God-like image would be restored, qualifying us all for a
return ticket to Paradise. That lost wisdom Roger Bacon
identified as the "mechanical arts" which, he pointed out,
had already been partly recovered. He preached the complete "unfolding of the divine wisdom through learning
and art" that would reunite us with God and Christ, and
then we too would become as Gods.

Fast forward from the 13th century to the 17th, still in
England and another Bacon—Francis Bacon, Lord Chan-
cellor and confidante of the King. We're only 40 years
from the Founding of the Royal Society, and Francis Bacon is compiling what will become its research program.
No magic here, no talismans, no incantation, instead a
detailed program for how to gather information and how
to use experiment to choose between alternative theories.
But then, towards the end comes this:

For by the Fall man declined from the state of innocence
and from his kingdom over the creatures. Both things can

be repaired even in this life to some extent, the former by
religion and faith, the latter by the arts and sciences.

The arts and science will restore to man the God-like power
over nature Adam enjoyed in the Garden of Eden. Francis
Bacon's book became a bible for scientists at the universities of Cambridge and particularly at Oxford who in 1662
were brought together to serve the King's pleasure in the
Royal Society.

The King of England was actually persuaded to lend his
name to the Society by a prominent member of a secret
organization. This gentleman wrote a constitution for the
Society and chaired its early meetings where, borrowing
from the secret organization's rules, he forbade discussion
of politics and religion. Members of this organization figured prominently in the new society.

The influence of this secret organization, famed for its
occult wisdom, had grown mightily when British noblemen
with an interest in the occult arts begged for admittance.
The organization was Freemasonry, and these outsiders
were admitted under the name of Accepted Masons. By the
time of Francis Bacon, Freemasonry tradition required a
high-ranking British aristocrat as its figurehead, which it's
had ever since.

And what was Freemasonry's secret? The practice of
a lot of pagan magic, but in the service of Roger Bacon's
apocalyptic vision. I found clues to this in a recent account
of FreeMasonry wisdom. FreeMasonry's purpose is "the
quest and recovery of something it has lost, but which by
its own industry it hopes to find." It is the loss of a word,
or rather of The Word, the Divine Logos, or basic root and
essence of our own being....It is in the cosmic parable of
Adam when extruded from Eden...

So three and a half centuries ago, in England, the two esoteric traditions of FreeMasonry and Puritan worship through the Mechanical Arts came together in the formation of a society "For the glory of God and the usefulness of man," to quote its official motto, its program based on Francis Bacon's plans for a new science. This explains why, at the same time as he was probing into the secrets of mechanics, dynamics and optics just as Francis Bacon had suggested, Newton was also studying the Book of Daniel for clues to the date of the anticipated Rapture. Many members of the Royal Society would have seen little difference between the two pursuits. Both were aspects of the plan they all subscribed to: saving the human race by recovering Adam's wisdom in preparation for the Rapture.

How long did this go on? Freemasonry and the Royal Society didn't separate until around 1850, under the direction of the Duke of Sussex who was both President of the Royal Society and Grand Master of the Freemasons. In other words, that relationship persisted into the same decade as the publication of Darwin's *On the Origin of Species*. Darwin's grandfather Erasmus Darwin who wrote grand epics on the subject of evolution was initiated into Freemasonry and Charles was very conscious of following in his footsteps.

Child of Positivism

150 years ago, just as science was shaking off its original millennialism, a new religion was springing up to take its place.

The object of worship in this new religion was no longer the Christian God, it was "The Great Being"—humanity, all of it; past, present and future. Worship consisted of private devotions, public ceremonies and sacraments administered by clergy. The year was divided into months of 28 days—13 of them, all starting on a Monday, named after great figures from history—plus an extra day. Each morning you sank to your knees, silently recalling women precious to you, preferably your mother, wife and daughter representing past, present and future, then you adored each of them in hysterical bursts of emotion. Each year you joined in 84 festivals celebrating one or another aspect of the Great Being. Nine sacraments administered by priests included the celebration of birth and death. Seven years after death came the last Sacrament: a public judgment by the priesthood on the memory of the deceased. If found worthy, he was incorporated with the Grand Being and his remains were transferred to a sacred wood adjoining the temple of Humanity. You saluted one another by touching the three organs corresponding to three sacred principles (one of which was love). Men had complete power in the family, women were forbidden to work except at educating their children. There was no divorce or remarriage even for the widowed.

That was "Positivism." But when scientists say they practice Positivism, it's not that aspect of Positivism they're referring to. They're referring to a hefty set of volumes setting down principles for the practice of science that would lead to endless technological progress.

Positivism was the life-work of a Frenchman called Auguste Comte. Where did Comte go for his inspiration? Back to Francis Bacon. To Roger Bacon. And Joachim. Like Joachim he divided human history into three stages. First medievalism: knowledge based on sacred texts. Then the Enlightenment, a stage of dry philosophizing. Finally Positivism, the sound application of scientific principles and practices leading to never-ending "Progress." The Age of Positivism was to be an age of knowledge based only on facts ascertained by measurement and experiment and filtered through mathematics. Nothing else any longer qualified as science.

In Positivist science, mind was outlawed. Comte forbade the study of subjective consciousness, referring people instead to phrenology. Only laws derived mathematically from scientific facts counted as valid wisdom. According to Comte:

> the positive [stage] is destined finally to prevail, by the universal recognition that all phaenomena without exception are governed by invariable laws, with which no volitions, either natural or supernatural, interfere.

In other words, conscious decision-making—what's referred to above as "natural volition"—cannot be allowed to "interfere" with Positivist science's invariable laws. To be scientific you had to account for everything in terms of only the application of mathematics to mass, distance, energy and time.

Stunning advances in physics in the early 20th century seemed to confirm the truth of Positivism. It spread from physics and chemistry into all the sciences, including biology. In 1949 it even spread to psychology. The mind (and our conscious self) became nothing more than a "ghost in the machine."

Today, Positivism appears to be losing ground. There's a movement in philosophy towards "post-Positivism," and in biology there's talk about genes having mind-like attributes such as intentions and selfishness. But for the majority of scientists Positivism still reigns supreme. It's the primary inspiration behind Physicalism. And it prepared the ground for natural selection.

For more on the limitations Positivism imposed on science and how science got into trouble by stepping beyond those limits see the "Fine Print" Appendix, "Science at the crossroads," starting on page 101.

CHAPTER 3

At the core of the threat:
natural selection

Physicalism—"only physical processes can make things happen in the physical world"—does have some logic to it. But without the support of one particular theory it would be just a footnote in science. That one particular theory is Darwin's theory of natural selection.

According to an article in the *New York Times* (April 15, 2007) we've entered "The Age of Darwin." A new "grand narrative" has crept upon us willy-nilly, and is now all around. "Once the Bible shaped all conversation, then Marx, then Freud, but today Darwin is everywhere."

And not just in the US. According to *The Economist* (April 21, 2007) this is happening not only in the USA and Europe, but worldwide. Everyone, all over the world, is being taught Darwinian evolution. The promise is, as knowledge of natural selection spreads, we'll all be brought together into a new global culture based on science. For the first time in history, theories about human nature will be the same everywhere, because Darwin's ideas apply not only to the world of nature but to us as well, because we too evolved. Whatever is true of evolution must be true of us, all of us, body and conscious self.

But suppose Darwin was wrong. Suppose that, in following him, we're wrong too. The result could be a worldwide catastrophe. If we're wrong about what drives evolution, we'd be wrong—all of us—about human nature.

Right upfront, I want to make one thing perfectly clear. It's not evolution I'm warning you against. We're evolved, I've no doubt about that. What I'm criticizing is not evolution itself, but the theory Charles Darwin came up with for what drives it—natural selection.

Natural selection is a purely physical process. If that's what really drives evolution, what really made us, then conscious decision-making can't exist. As a result, natural selection has become the main battleground for the fight for and against the conscious self.

Natural selection isn't just a scientific theory. 150 years ago it liberated people from oppressive religious doctrines. But today it, in turn, threatens to become our oppressor. As our primary origin story it sets limits on what we can expect of our selves. We can't be more than whatever it can create. As part of science's toolkit it wields enormous authority. If it's wrong, it's vital we find out. If natural selection turns out to be false, if you can't any longer "prove" we evolved through purely physical processes, then the self escapes Physicalism's deadly embrace.

Is resistance futile?

Maybe you're thinking resistance to natural selection is pointless. Could evolutionary science be wrong about consciousness?

Sure. It's been wrong before. In the 19th century evolutionists preached, "Let the strongest survive." We didn't like that, so we stopped them—we insisted on concern for the weak as well, and together we've all become stronger. A century ago evolutionists said people who were "less fit" should be sterilized so their genes wouldn't "contaminate" the gene pool. We didn't like that and again we stopped them.

Now they're attacking consciousness. We don't have to put up with that either. We don't have to give up believing in conscious decision-making and personal responsibility just because of what science says. We mustn't shrink from peering over the fence at what professional evolutionists are doing and criticizing them when we think they're wrong. They have been wrong before, and they'll be wrong again. If we think they're wrong about consciousness we should say so.

Let's take a closer look at the theory of natural selection. Here's how Darwin summarized it in *The Origin of Species*:

> "Can the principle of selection, which we have seen is so potent in the hands of man, apply in nature? I think we shall see that it can act most effectually....can we doubt (remembering that many more individuals are born than can possibly survive) that individuals having any advantage, however slight, over others, would have the best chance of surviving and procreating their kind? On the other hand, we may feel sure that any variation in the least degree injurious would be rigidly destroyed. This preservation of favorable variations and the rejection of injurious variations, I call Natural Selection."

Here's his theory boiled down to bullet points:

1.   Creatures that succeed in surviving to reproduce pass the characteristics that made them successful on to their descendents.

2.   As the characteristics making for success accumulate
in the species, it becomes more adapted to its surroundings,
eventually leading to the formation of new species.

Seem obvious enough? Well, I disagree with it, and for
two reasons. First, it's just bad science, as I'll show you in
Chapter 3. But also, it can't account for the one thing we
experience directly—conscious decision-making. For that
we need an entirely different kind of science.

How I came to doubt natural selection

I didn't start out doubting Darwin's theory. In fact I was
one of Darwin's most enthusiastic admirers. But I started
doubting his theory when I realized it didn't tell me anything useful about evolution. Here are some examples:

1. I'd be enjoying playing with my cat, then I'd wonder
what that said about me. Was I so desperate for company
that I'd do whatever a creature of some other species wanted me to do? Was I letting the cat make a fool of me, making me pretend to be another cat? Or I'd be in the garden
and I'd be amazed at something a plant was doing, how it
was growing, and I'd want to deepen my appreciation of it.
But natural selection didn't provide me with answers. Instead, it acted like a wet blanket, dousing my interest. The
answer to everything was, it just evolved that way, through
natural selection. Nothing was explained. That seemed
odd for a scientific theory. Usually scientific theories illu-
minate. There's something wrong with a theory that does
the opposite, blunting your appreciation for whatever you
want to know more about.

2. A friend told me there was an epidemic of suicide
among young men 16 to 23 years old. Apparently it was
a new intensification of adolescent existential angst, and

parents had to keep careful watch over their sons. I tried to think of a way to relieve this kind of morbid thinking. But us having evolved through natural selection provided no basis for a message of uplift or hope. Our culture seems blighted by its new origin story, and that didn't seem right.

3. Why does so much of our present thinking come from the past? Mind and will from the Greeks. Sentimental feelings such as passion and pity from Christianity. Humanism from the Renaissance. A greater sense of personal responsibility from the Reformation. Individual rights from the Enlightenment. But now? Nothing. Instead of the main message of our time—natural selection—making our culture richer, it seems to be hollowing it out.

4. Whenever I looked at natural selection for whatever it was that made evolution so creative, that I could apply to enhance my own creativity and intelligence, I came up empty-handed. Natural selection failed to account for creativity, for feelings as the cause of behavior, and any motives other than sex and survival. Though supposedly it had made me, it told me nothing I could use to improve my self. Odd!

But what finally made me challenge natural selection was the logic behind it that denied free will—"Natural selection is a purely physical process that can act only on materials, so brains can evolve but consciousness can't. Because consciousness can't evolve it can't exist. What actually makes decisions for you is not your consciousness, it's chemical reactions in your brain. Your conscious thoughts are just a byproduct of those chemical reactions. Consciousness can't act back on your brain to make anything happen, any more than your shadow can."

And "Since what you do gets decided for you in your

brain chemistry before you even become aware of it, you really aren't responsible for what you do." It's that old-time Physicalism, now "proved" by the theory of natural selection.

That's why my attack on Physicalism takes the form of an attack on natural selection. In Part 3 I show you how.

CHAPTER 4

The "Violent Engine" hypothesis

To give us something to compare natural selection to I needed a name for what "really" drives evolution. One day, as I was idly playing around with the letters in the words "meaning & evolution" I noticed that by rearranging the letters in the word "evolution" I could make the word "violent." Rearranging the letters in "meaning" I came up with "ngine." Then I realized I could use the "e" of "etc"—the "&" between the two words—and make "Engine." "Violent Engine." And in the remaining letters I found the Latin words for "I love" and "you."

"Violent Engine, I love you." What a great name for whatever it was that first created and then drove extinct 99% of the creatures that ever lived! And I do love it, because it made me and all the other wonderful creatures I share this Earth with.

I'm going to use "Violent Engine" as a name for what really drives evolution. We don't know what it is, but that doesn't matter. It's whatever's brought living creatures to where they are today. Is natural selection the Violent Engine? Could natural selection pick up the reins from the Violent Engine and take over?

Let's begin by asking, what does the Violent Engine— the real mechanism behind evolution—do? It creates and maintains living creatures. How is a living creature different from a pile of rocks in a puddle? It's alive. And how can you judge what that means? By what happens when that creature dies—within a few days it putrefies and becomes just another pile of rocks (bones) in a puddle. Or, to put that in more scientific language, it "returns to equilibrium with its environment."

So one thing the Violent Engines does is stop a living creature, while it's still alive, from becoming a pile of bones in a puddle. It keeps living creatures far, far away from equilibrium with the environment.

Another thing it does is, over time it makes new creatures that are even more complex, that it has to maintain even further from equilibrium. Put in more scientific-sounding language, it "generates an increase in ordered complexity."

Whatever the Violent Engine is, we know it can "both generate, and then maintain, an increase in ordered complexity." That's the challenge natural selection has to meet.

CHAPTER 5

Natural selection fails
the "Violent Engine" test

Natural selection—child of Positivist science

A few years after he returned from his trip around the world the young Charles Darwin came across a preliminary outline of Auguste Comte's Positivism and reported himself "enthralled" by it. He returned a few days later and reread it so intently, he wrote, that he developed a headache from the stress. Positivism became a major influence on Darwin at just the time he was working out, in his famous secret notebooks, how natural selection worked.

But he didn't publish his theory. He might never have published it if the young naturalist Alfred Wallace hadn't submitted for publication almost exactly the same theory. By gentleman's agreement, they published simultaneously.

Why did Darwin hesitate so long? He didn't know if the idea was any good. He confided it to one or two friends, begging them to judge it, but they wouldn't commit themselves. In 1844 a series of books titled *Vestiges of the Natural History of Creation* described in detail how all living creatures evolved from earlier creatures. Alfred Wallace thought those books told you everything you needed to know about evolution except for the mechanism driving it.

Evolution itself was out of the bag, the storms generated by it were raging over someone else's head. But still Darwin kept his notes on his mechanism hidden away. Altogether he kept his theory of natural selection under lock and key for 20 years.

Darwin was an admirable person and a great scientist. He was a genius naturalist and observer, and a very good experimenter. But recent books hint that he wasn't a great thinker. He was no Newton. When later he came up with a theory he called "pangenesis" with "gemmules," what we call "genes," experiments showed they couldn't be spread throughout the body the way he imagined. But he couldn't refine his theory by locating his gemmules in the cell nu cleus where they would later be found. When Wallace confronted Darwin with why natural selection couldn't account for civilized talents in humans, Darwin wouldn't discuss it, he merely begged Wallace not to destroy "our child." In *On the Origin of Species...* though it was natural selection that got most of the attention Darwin actually described several different mechanisms for how evolution acts, one of them ("use inheritance") very much like a discredited theory proposed half a century earlier ("Lamarkism"). And the idea behind natural selection wasn't that hard to come up with. Two other people had published it many years before he and Wallace thought of it. And the inventor of the phrase, "survival of the fittest," Herbert Spencer, had come up with something similar around 1850.

So why, besides for being an all-round wonderful person, is Darwin so celebrated?

Darwin became the figurehead of a movement that started in earnest in 1850, dedicated to breaking the Church of England's monopoly over admission to the universities and the professions. Darwin was a famous and respected naturalist. He'd put his name to a theory that snatched from the Church its primary source of authority—its claim to represent the Creator of all living species. Shortly after publication of "The Origin of Species..." Darwin's principal supporter insulted a Church of England bishop in public, starting the tradition of "Darwin's Bulldogs," enthusiasts for natural selection who feel obliged to savagely attack anyone who dares to challenge it.

By the time Darwin died this movement had become so powerful that, in just a couple of days, it was able to swing national opinion behind Darwin being buried among the greats in Westminster Abbey. Earls of the realm were among his pallbearers. The movement honored its champion. And by honoring him it added enormously to its own influence and power.

50 years later another movement wanting to boost its influence and power turned to Darwin. By the 1930's all the other sciences had come to envy physics for its success in applying Positivist principles. Biologists, too, wanted to gain respect by making their field Positivist. Dragging Darwin out of obscurity, they made him once again a figurehead. Myths were created around him. To conceal his failure to properly locate the gene in the cell nucleus, a great deal was made of Gregor Mendel's work with peas, as if only that could have led to the right theory.

Darwin has remained evolution's patron saint to this day. In fact, if he sprang to life today, he'd have a hard time discussing evolution with modern evolutionists. They'd be respectful I'm sure, but point out how lay people like him can't talk about evolution, it's all a matter of statistics now that lay people like Darwin can't understand. Actually though, beneath all their advanced theorizing, the engine driving everything is still the same, the simple theory that Darwin's celebrated for—natural selection.

What the Violent Engine does

What does it take to keep a living creature alive? If you think about it, it must involve millions of things. How the kidney tubules work. How nerves carry signals around the heart to make it pump just right. How lungs work. The eyes. The ears. The feet. Every biochemical reaction. I could go on and on. Altogether the most fantastic piece of engineering you could possibly imagine, all built just right to work together for a lifetime without going back into the shop. How many things have to be just right? A million? Probably billions.

Not only that, but whatever's no longer needed gets lost through what Darwin called "disuse," like our tails and most of our appendix, like creatures that lose their eyesight once they start living in dark caves. And new useful features appear.

The Violent Engine obviously can do all this, because it happens. Can natural selection do it?

Could natural selection be the Violent Engine?

If you came up with natural selection today as the mechanism responsible for evolution I think you'd have a very hard time getting people to accept it. But once everyone takes it for granted, as they do today, it's just as hard to explain what's wrong with it. But that's what I'm going to do. I'm going to explain why natural selection can't be the Violent Engine, the mechanism responsible for evolution.

Here's what we want to account for—creatures being kept alive far from equilibrium, all the features they need being maintained while features they don't need get lost. And every so often new features appear.

Here's what natural selection does—and all it can do. In each generation it divides a population into two: into those creatures that reproduce, and those that don't. That's it. For each creature: pass or fail. For the species as a whole, those with something, that do get to reproduce, and those without it, that don't. Natural selection can make only that one distinction. Could that be enough to drive evolution?

Here's a parallel. Remember the story of the Dutch boy who saved his village by using his finger to plug a leaking hole in a dike? A dike or levee keeps the water behind it high above ground level, just as the Violent Engine keeps a living creature far "above" chemical equilibrium with its surroundings. The Dutch boy's finger is like natural selection, it can prevent only one "leak" at a time. Now imagine the levee itself suddenly disappearing. How many fingers will the Dutch boy need now, to hold back all the water? Billions of fingers, and a lot more strength than he's got. That's like natural selection trying to keep a living creature up in the air, far from equilibrium with the environment. It's got only one "finger"; it can make only one distinction, pass or fail, one choice. What will that be? Kidney tubules working properly? The heart's nerves being routed correctly? The lens in the eye working? Any other one of a million or billion other factors? Natural selection can't select for them all at once. It's just not equipped to do the job.

An evolutionist friend of mine said "Just group together all the factors needed for keeping a creature alive, call that 'fitness,' and select for just that as a single item, in one go." But a living creature doesn't work like that. Each factor essential for staying alive has to be selected for separately, needs its own "finger." Otherwise you couldn't lose just the things you didn't need any more, while keeping those you did. A land creature taking to the water to become a whale couldn't lose features needed on land but worse than useless in the water, like legs. No, all these things have to be selected for separately, all at the same time, some being kept, some being lost.

Here's another parallel. You could say, keeping living creatures so far from equilibrium with their surroundings is like preserving them in a deep freeze, even while they reproduce and evolve.

Now natural selection comes along and says, "I know how to do this. I'll take over." The creatures are taken out of the Violent Engine's deep freeze and natural selection busies itself selecting them when they reproduce to keep them "evolving." It divides them into two piles, keeping one, discarding the other. But as it works the creatures start melting and in a short time there's nothing to see but puddles and bones stretching away in all directions. Being able to make a single distinction isn't enough to keep those creatures from "melting" (dying). Natural selection doesn't have the necessary horsepower.

Modern evolutionary math doesn't work

Darwin understood all this. He knew natural selection had to account for maintaining existing features, while creating new ones, as well as losing those you no longer wanted. So how come modern evolutionists don't?

In 1930 a guy called Fisher decided to "modernize" natural selection by turning it into statistics and ever since then evolutionists have been followers not of Darwin but of Fisher. Fisher made them look more scientific by relying on statistics.

But while Darwin was a gentleman, showing up with a full meal of entrée and dessert, Fisher just brought dessert, and evolutionists have been making do with dessert ever since. Fisher dealt only with how new features could evolve, one gene at a time. That's much easier. But now you're no longer accounting for what keeps all a creature's existing features from disappearing, for what keeps it from collapsing into equilibrium with its environment until life ends (those bones in a puddle). Natural selection can't do what's needed to preserve existing adaptations, let alone create new ones. And if evolution actually involves changes to whole clusters of genes rather than to individual genes, all Fisher's theorizing collapses.

Don't just take my word for this. In "Evolutionary Statistics" in the "Fine Print" Appendix on page 106 you'll find some of evolutionary theory's heaviest hitters giving you the inside dope on evolutionary statistics and showing you why they don't work.

CHAPTER 6

Collapse of the "epicycles"

I've shown you that natural selection doesn't work. But natural selection is like a hydra—as soon as you dispatch one theory, another one pops up in its place. You have to dispatch them all.

"Epicycles"

There's the name you give new theories introduced to make up for flaws in an old theory. When people thought the Earth stood still and the sun, planets and stars all went round it at different speeds, they also assumed those bodies all moved in perfect circles. But to keep the movements of those bodies in line with appearances—to "save the appearances"—they had to keep adding additional smaller and smaller circles to the system—"epicycles"—until they ended up with the "heavenly" bodies moving in over 50 circles great and small! Then someone proposed the Earth rotates and goes round the Sun, and all the epicycles disappeared in a puff of smoke.

Maybe one day all natural selection's "epicycles" will also disappear in a puff of smoke. But for now, each one of them is claimed to be an independent theory in its own right, and the more of them there are the more that's supposed to confirm natural selection. That's like saying, the invention of more and more epicycles just goes to prove the

Earth stands still! After all, if it didn't, scientists wouldn't still be "discovering" them!

Let's turn to natural selection's primary epicycle, "mutation."

Variation and "Mutation"

What's most amazing about evolution is its creativity—its ability to create new kinds of living creatures. Along with new creatures come new genes. So if you take a gene-centered look at evolution you find that new or more complex living creatures go along with new order in the genes.

"Mutation" is the opposite. It's taking genes that have already evolved and replacing them with genes changed at random. That could be by physical damage, could be by genes not sorting properly when cells divide. It's loss of some of the order among genes that evolution created. It's one tiny step down from life towards the bones in a puddle. In other words, it's injury.

Mutation is a very bad idea. It's a would-be solution to a non-existent problem. At least, there isn't a problem until you introduce natural selection.

The problem has to do with "variation." Nature ticks along very nicely with lots of genetic variation, so variation by itself isn't a problem. Wonderful new variations pop up all the time, and old ones drop by the wayside.

Fine. There's only one problem. Someone comes up with natural selection as an explanation for how evolution works. But natural selection works by eliminating the less fit which reduces variation. Since reducing variation is the very opposite of what you want, variation has now been turned into a problem. What do you have to do to your theory of natural selection to give the appearance of variation actually increasing?

What evolutionists came up with was a second process to put variation back—"mutation." We'll inflict damage on some creatures' genes, at random, reverting them partway back to the bones-in-the-puddle equilibrium. We've restored variation.

Let's see how that works out in practice. We'll pit one species evolving through the Violent Engine against a second species evolving through natural selection plus mutation. In this second species, natural selection first reduces existing variation. Then mutation injures the rest.

How long shall we give this contest!

You can't have it both ways. Either there's a process in creatures for making new variation, or there isn't. If there is, it doesn't need damage to make variation from. If there isn't such a process, then damage can't be made into new variation. Either way, you can't use "mutation" to explain where variation comes from.

Mutation may be a bad idea, but it's very hard to disprove. Whatever's responsible for variation, it's bound to involve changes in genes. Suppose they're not random damage. How are scientists to tell the difference? They'll call any change in genes "evidence for mutation, which proves natural selection."

"Mutation" isn't a proof of anything, except that the theory of natural selection needs an awful lot of epicycles.

Turn to "Disposing of more 'epicycles'" in the "Fine Print" Appendix starting on page 111 for challenges to:

Competition

"Kin selection"

Why selection isn't important for evolution

Who's clustering the genes?

The origin of species, still a mystery
Wallace, co-discover of natural selection, turns
against it

I also reveal the laboratory where I do my research and where all my ideas come from.

CHAPTER 7

As natural selection fails, Physicalism fails

Will natural selection follow "The Blank Slate" into oblivion?

Could a theory that's been accepted as scientific for 150 years possibly be wrong? It could, that's happened before, and to a theory that lasted over 300 years. It's the theory of the blank slate, or Associationism, and it's been abandoned only recently, in the past few decades. I say more about it in Part 2.

Why does a theory like Associationism fall out of favor? Like species, some theories seem to just go extinct. They keel over and die. A once-trusted theory suddenly seems like a dry husk that's keeping us from growing, and people look for something else. People start to question the assumptions behind it, and find it doesn't make as much sense as it used to. I believe the same is about to happen with natural selection.

Once natural selection goes, Physicalism will fail. Without a purely physical mechanism for evolution, Physicalism would once again become just a minor historical curiosity. The self would be saved.

What might a non-purely-physical mechanism of evolution look like? In Part 2 I draw up a blueprint for just such a mechanism.

Part 2
Blueprint for a new self-based science

CHAPTER 1

Dubious assumptions
well worth challenging

To come up with a new theory to replace natural selection, what do we need to do first? We need to take the blinkers off. We need to identify the assumptions behind science and natural selection that make it so hard to consider alternatives, and set them aside. If we don't, like invisible chains those assumptions will hold us back. We need to start by cleaning house.

I start in the basement, with the religious foundations of science and the assumptions they built into it. I believe those assumptions still influence how science operates and how we all think. Let's sweep those cobwebs away.

ASSUMPTIONS ABOUT HUMAN NATURE

Once religion got science up and running, science took off and grew, more and more along the lines of pure reason. Scientists today think of themselves as motivated only by logic and reason. Are they right? Or are there subtle biases built deep into what they study?

Human nature's not worth studying

To Millennialists like early members of The Royal Society, human nature is sinful, fallen. So why study it? First, that won't help you get to Heaven. Second, the Fall made human nature distasteful to God so we'd better not identify with it too much, or devote too much attention to it. And anyway it's only temporary. Come the apocalypse our original perfect image will take its place.

Could that be why the "hard" sciences still tend to ignore human nature? If so, this is one assumption I think it's very important to drop. New assumption:

Let's welcome human nature in our new science.

Human exceptionalism

When something gets shouted from the housetops you can't really call it an assumption. An assumption is usually something you take so much for granted you don't even talk about it. So when evolutionists insist, absolutely INSIST, there's no intelligence behind evolution, you can't really call it an assumption.

You have to wonder, though, why they insist so much on something they can't prove; that something *doesn't* exist, like there being no intelligence in evolution.

Of course, they say we humans have intelligence. But only us, not the rest of nature. Doesn't that echo another part of the apocalyptic message—we, and only we, will be found worthy of being carried up to heaven?

Claiming we're different from all other creatures is called "human exceptionalism." In science that's usually a no-no. But that's exactly the error science falls into when it grants creativity and intelligence to humans but denies them to other evolved creatures and to the processes of evolution itself. Let's abandon this assumption.

Let's not deny intelligence to other living creatures and the processes of evolution.

A self without consciousness

Did you know that 150 years before the French Revolution, in fact just before the founding of the Royal Society, the English had a long and bloody revolution of their own? Religious Puritans fought the King, won, cut off his head, and for ten years ruled the country. Then Parliament decided it had had enough of government by religious "enthusiasts" and brought in another King.

It was a revolutionary time. And out of it, along with the Royal Society, came a revolutionary new theory of human nature. We weren't created in God's image, it said. In fact we don't come with any image at all—we come into the world with our minds a blank slate. All we have are the abilities to make associations between things, and the experience of pleasure. When two things enter our attention at the same time and we feel pleasure, that makes a permanent association in our minds. That's where minds come from. No God. No soul. No consciousness. Just these associations.

This theory, called "Associationism," became science's model of human nature—a tangle of associations with self, soul and consciousness vacuumed out of it.

But then science leaves out just those things most of us most want to know more about—our selves, human nature.

Let's abandon the assumption we're nothing but a tangle of associations.

The self as adaptation

The theory that our minds are nothing but a tangle of associations set the stage for natural selection. Besides the associations we make on our own there are others we all share. Where do they come from? They seem to come from all of us being members of the same species. That opened the door for Darwin to suggest that those associations came from whatever it is that makes species, which according to him was the purely physical mechanism of natural selection.

What happens when we make associations? We feel emotions and we make faces. In a book on facial expressions Darwin showed we make just the same kinds of faces as other animals. Facial expressions, in other words, are just biological "adaptations." And so, probably, are the emotions that go along with them! What to us is consciousness is actually nothing more than what makes animals make faces. If they're not conscious, there's no reason to assume we are!

Let's drop that assumption. In fact, let's turn it around. If we have conscious experience when we make faces, maybe animals do too. Maybe conscious experience is quite common in nature.

Let's not set any limits to who or what is conscious.

EMINENT VICTORIAN ASSUMPTIONS

The first edition of Darwin's On the *Origin of Species...* sold out the day it was published. It created a sensation. In just a decade his theory of natural selection became widely accepted.

If natural selection seemed so obvious to Darwin's Victorian followers, why didn't someone come up with it earlier? I think because the Victorians developed new assumptions of their own. People before that, who didn't make these assumptions, would have thought natural selection was crazy and rejected it.

How about us? If we came face to face with natural selection for the first time today, would we accept the assumptions behind it? Or, looking at that the other way round, if we accept natural selection, does that force us to accept those same assumptions? Does defending natural selection shackle science with a web of obsolete assumptions it can't throw off, that stop it from exploring in new directions?

What are the assumptions lying behind natural selection?

"Evolution"

There's an assumption built into the word "evolution." And we get that assumption from the Victorians.

Before the Victorians, most people took it for granted that all living creatures were created by God. Those creatures were like us except God gave us reason and souls as well. But in Victorian times belief in God began to weaken. Even if God did exist, people imagined Him as having started the world off and then left it to care of itself.

If that was so, where had all the living species come from? The Victorians became obsessed with this question. They embarked on a quest for the "origin of species" and began sending naturalists fanning out across the world. One of these was Charles Darwin. And Darwin came up with their answer—natural selection.

I say there's an assumption built into the very word "evolution" because, since Darwin's day, we've encountered other mysteries just as profound. How living creatures develop from a single egg all the way to a full adult. And how, at every stage along that path, a creature sustains itself, keeping itself from collapsing into equilibrium with the environment (becoming bones in a sticky puddle). These two mysteries go by the names "development" and "homeostasis." So for us, the great question is how life maintains order within itself, and then goes on to create new order, either during development or during evolution. Evolution ceases to be something to account for alone, it becomes part of something much more complicated, what I've been calling the "Violent Engine."

This is one very significant way natural selection stops science from moving on. It reinforces the Victorians' assumption that you can separate evolution out as something to account for by itself.

Let's not separate evolution out from development and homeostasis. Let's try to account for them together.

Atomism

Very early on, modern science divided up between what you could account for in terms of atoms, and what you couldn't. As it turned out, accounting for things in terms of atoms led to runaway success, and science found itself with a lot of tools and theories involving atoms. There's an old saying, if all you have is a hammer, everything looks like a nail. If all your theories involve atoms, everything looks like atoms. By Victorian times, everything-is-atoms had become a common assumption—if you couldn't account for something in terms of atoms you didn't really understand it. If you were a Victorian and what you wanted to account for was evolution, the first thing you'd do was look for its atoms. And one by one the Victorians found them. First, in variation. They divided variation up into a new kind of "atom."

Before that, up to about 1750, people thought of each living creature as a unique individual of a particular kind. Each had been created by God, complete and perfect, just the way He wanted it. You could list ways creatures differed, but you'd assume that was because of the words you had to use. If you said one is "bigger" than the others you wouldn't have meant it inherited "bigness" as a separate "characteristic." But the Victorians did. "Characteristics" were their new "atoms."

Having divided variation up into "characteristics" Victorians like Darwin then assumed these new "atoms" of variation got inherited one by one independently of each other. Darwin probably assumed this because it's how people thought about artificial selection—breeders selected for particular "characteristics" like disease resistance or strength.

"Bigness" now became a characteristic that creatures could inherit separately. Instead of seeing living creatures as all unique individuals you could think of them as all the same except for differences they inherited in a limited set of characteristics.

Once the Victorians had got used to thinking in terms of "characteristics" they began looking in the body for the "atoms" that coded for these "characteristics." Darwin referred to them as "gemmules" as I said before. He pictured them as particles about the size of bacteria floating freely throughout the body so whenever cells divided there was always a complete set of these particles in every cell. Later, scientists found his gemmules strung together on chromosomes in the cell nucleus, and renamed them "genes."

Scientists still tend to think of living creatures like that. Each individual in a species is a standard-issue biochemical factory, pretty much the same in all individuals, plus a set of characteristics that get sorted at random when those creatures reproduce and account for what makes them different. There are now machines that can read the entire genome from end to end. That's a mighty big "hammer." So now scientists "know" exactly what variation consists of—all the genes from one end of the genome to the other.

Should we continue to make this assumption, that everything comes in atoms? Sure, some inheritance comes in particles—the information for making proteins, for example. But does it all come that way? Here's an analogy—a magician on tour gets sick. He's sent his bag of tricks on ahead, but he can't make the trip, so he calls up and wants to cancel. But the theater manager says, your bag of tricks is here, we'll just put it on stage. Isn't that where the magic is? Well, no. You can't do magic without it, but the equipment isn't enough by itself. Similarly, for life to go on, proteins have to be specified precisely, atom by atom. But when you've found all the "genes" that code for those proteins, you haven't necessarily identified all the "magic" it takes to run a living creature.

Today, in genetics, atomism still reigns. Scientists claim they've identified around 30,000 genes in the human genome. The rest appears to them to be "junk." Those 30,000 "atoms" presumably code for around 30,000 "characteristics." That's certainly a lot, and might have impressed the Victorians. But is it enough? The body has something like 250 different kinds of cell, over 20 different organs, 600 muscles and 200 bones, over 1000 entities altogether. 30,000 "characteristics" doesn't seem nearly enough to me, especially since all these body parts have to be defended against failure not only in their final state in adulthood but also through every instant during development. We've found the magician's box of tricks, but we're missing some of the magic. Maybe we need to look somewhere else. But when you've been picturing nature in terms of atoms for a couple of centuries, how do you turn around and imagine it any other way? Every way you try thinking about it ends up carrying you back to atoms.

Here's how I think about the genome: it's a hologram. What scientists call "phenotypes," such as different breeds of dog, are pictures you see in the genome as you view it from different angles. I prefer that to imagining those breeds as random assortments of "atomic" characteristics.

Inheritance doesn't necessarily come in atoms that can be sorted independently.

Adaptation and variation

What's special about species? To the Victorians it was how well they were adapted to their surroundings. If God wasn't responsible for that, how else could you account for it?

Darwin's answer, of course, was natural selection. Natural selection automatically concentrates in offspring those "characteristics" that help creatures survive to reproduce. In other words, that species' creatures become better and better adapted. Hey presto, the primary problem—adaptation—has been solved.

Of course that created the secondary problem of providing the variation natural selection needed to work on, so scientists came up with the idea of mutation. But for the Victorians finding a source of variation was a secondary problem. The primary problem was, what makes creatures so well adapted to their surroundings.

Suppose, though, you turn that assumption around. Think about all the different breeds of dogs and how much they vary. Apparently, before breeders separated them out into different breeds, all the characteristics of those different breeds existed together in one species of wild dog or wolf. That's a huge amount of variation. Maybe that's what's most distinctive about living creatures, what most needs accounting for—not how well they're adapted to their environment but how much variation there is among them. Maybe, when there's enough variation, adaptation follows automatically. Maybe the main issues for biology today should be dealt with in a book with the title *The Origin of Variation* that doesn't mention either adaptation or mutation.

Let's assume variation is at least as important as adaptation.

Quality control

The same decade Darwin published his *Origin of Species* the 1851 Great Exhibition opened in London, first of the great World Fairs. It revealed to an astonished world a totally new industry: mass-produced steel machinery powered by steam engines. What made this new industry possible was the invention earlier in the Victorian period of machine-tools, which in turn made it possible to manufacture machines out of interchangeable parts.

Maybe from having grown up with this industry, Victorians like Darwin seem totally un-fazed by the problem of making living creatures out of interchangeable parts— their new "characteristics." They seemed to assume living creatures came just like the parts of a steam engine, all interchangeable from one creature to another, that you could then attach "characteristics" to and expect them all to work together just fine, in any combination.

Maybe they came to this assumption because of how successful "artificial selection" is, where you select for the same "characteristics" over and over again in successive generations. That's how breeders separate out new "breeds" like the various breeds of dogs. Darwin thought nature worked the same way. He assumed characteristics got sorted at random in each generation, then nature selected for the combinations that "worked" best.

But nature doesn't work that way. Think about what happens in the wild and you'll see that characters don't sort at random. Take all the characteristics in all the different breeds of dogs you know and imagine them all packed back into the genome of the original wild breed. Now imagine those characteristics all being sorted at random to make new puppies! Think how those puppies would vary! Some would come out like Great Danes, some would be like pekinese, some would be like poodles! Yet in nature, wild puppies don't vary that much. Think of a pack of wolves, how alike they are. Adults wolves are much more similar than they "need" to be just to survive. They don't look like the result of a random selection of characteristics. So let's not treat how living creatures vary as "atoms" of inheritance that get sorted at random.

Let's not assume evolution works anything like the artificial selection breeders use.

Let's not assume characteristics are selected at random.

Predictability and creativity

Can we take off the Victorian "atoms" spectacles? Here are some things that resist being seen as atomic.

Skulls. I like skulls. I have skulls of a beaver, and of a dog. They're two very complicated, very different, sculptures. Can you break down their shapes into atoms? I can't imagine it. Can you imagine converting one into the other as if they were made of rubber you could reshape just by stretching? I don't think so. They're different creatures, and their skulls are fundamentally different, as different as people are.

Look at a cat's head. See how the hairs lie differently

around the eyes and the nose and the ears. Could you express that in atoms? I don't think so. And that's just how the hairs lie on its face!

From experiences like these, I've concluded that living creatures' characteristics don't divide up into atoms, "variation" isn't coded for by "particles" of inheritance. Variation is how creatures differ from each other totally. Not atom by atom.

Once you say that, the theory of natural selection collapses. It can account for evolution only in terms of atoms. In turn, mutation collapses—if variation doesn't come in atoms, as random changes in individual genes, it can't be selected for by natural selection.

Let's not assume the forms of living creatures divide up into atoms.

Science can't see selfs

All those stages science passed through—Millennialism, Associationism, Positivism, natural selection—notice one thing they all had in common? They all, for one reason or another, denied the self. The result is, science has become this fantastic searchlight you can point anywhere and it'll tell you what's going on. Except, it's blind to selfs. Are there any selfs in the world? We know there are. But if you ask science it'll say, No. If what you want to find out about involves other selfs like ours, it can't help you. So what's likely is, the most interesting things left to find out will be things that involve selfs. To me, that sounds like a good place to be looking.

Let's not be blind to selfs.

Behind every assumption, a question

The assumptions behind natural selection support and

build on each other. Look at how they follow one from another: First, atomism, then seeing variation in living creatures as separate "characteristics," leading to seeing living creatures as machines made of interchangeable parts. What looks like creation is actually just mutation—random variation in "characteristics"—that gets whittled down by natural selection into new ways of adapting. Finally, this combination of mutation and natural selection acts like a computer to "solve" challenges presented by the environment—adaptation.

Supposedly, none of this requires intelligence. What looks like intelligence in us, the argument goes, is really just genes for "strategies" that natural selection selects for, making us better adapted to our environment. We can't have minds because the process that made us, natural selection, can't make minds. If we think we have minds that's simply a muddled sensation of these mechanical strategies operating in our brain chemistry.

Before looking for alternatives to natural selection I have a suggestion: peel these assumptions away, one by one. Shed them all, and stand away. You have to look at them from a little distance and see them as a set of assumptions built up to support someone else's answer to what, for them, was the most important question. Having shed all those assumptions, you then peel away their answer to that question. Then you peel away their question. And you ask yourself, for someone today conscious of being conscious, what is the appropriate question?

Here's my question. What mechanism creates our conscious decision-making selves? Once we understand that mechanism, we will know our own selves much better, and be able to develop technologies for improving them.

In preparation for Chapter 2 I've identified some of the

assumptions behind natural selection, and made new assumptions out of them. Here are my new assumptions:

Let's welcome human nature into our new science.

Let's not assume the self is only a tangle of associations.

Let's not deny intelligence to other living creatures and to the processes of evolution.

Let's not set limits to who or what could be conscious.

Let's not be blind to other selfs in the world.

Let's not assume living creatures's forms come as atoms.

Let's not assume evolution works anything like the artificial selection breeders use.

Let's not assume evolution involves characteristics being selected at random.

Let's not focus more on adaptation than variation.

Let's not start out separating evolution off from development and homeostasis.

CHAPTER 2

More attributes of the "Violent Engine"

What does a theory of evolution have to account for? To me, it has to explain where new creatures and new variations come from. I'm going to call that "creative capacity."

Let's start by doing just the opposite of what Darwin did. Darwin's mentor Sedgwick didn't approve of how Darwin carried out his search for the mechanism behind evolution. Instead of looking at evolution as a whole and deducing his mechanism from that, Darwin came up with his mechanism first and then just piled up example after example to support it. Let's do what Sedgwick suggested, and look at evolution as a whole, from beginning to end, and see if that suggests what kind of mechanism drives it.

First, let's see how good a fit natural selection is.

How evolution's creative capacity changed over time

Natural selection is a very simple and crude mechanism. Survival is a simple pass-fail test of each creature. A creature either survives to reproduce, or it doesn't. You can't complicate that or "improve" it.

Assuming a process like natural selection has what I'm calling "creative capacity," it would have a fixed creative capacity, like this:

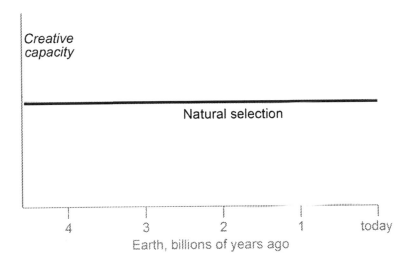

There's not much you can do to make such a simple process more creative. In fact, all you can do is make it less creative. And over the course of evolution, that's what happened. Check off these points and see if you agree with me that they'd each make natural selection less effective.

    -Creatures took longer to get to reproductive age so generations came further apart,

    -Creatures produced fewer offspring,

    -As creatures got larger, breeding populations got smaller,

    -As creatures got more and more adapted to their environment, that made it less likely random mutations would be improvements,

    -The more genes there are in the whole genome the more the effects of individual genes get diluted and the less difference between individual creatures there'll be to select for,

-As more features accumulate that selection has to maintain, the less selective capacity there'll be left for creating new features.

So here's how I imagine natural selection's creative capacity actually changing over time as evolution proceeded.

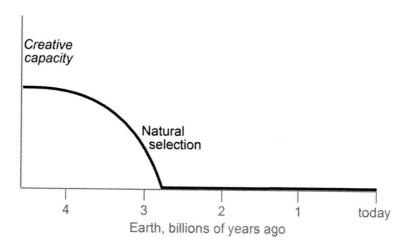

Search for a more plausible mechanism

Now let's compare that to how evolution actually changed. Evolution started out with only simple creatures, which continued to get more complicated as time went along. So the creative capacity of evolution, and presumably the mechanism driving it, actually changed more like this:

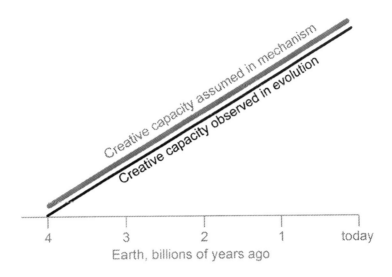

Obviously, whatever the mechanism behind evolution is, it isn't a mechanism like natural selection with a fixed creative capacity. Instead, it increases in creative capacity as evolution goes along. That suggests evolution generates its own creative capacity. The mechanism itself evolves.

How intelligence grew over time

Now let's look for a plausible origin for the conscious self. Since I don't know how to identify conscious selfs in any other creature but us, I'm going to look for a surrogate for the conscious self, something I can recognize: intelligence.

Here's a diagram showing stages in the evolution of intelligence.

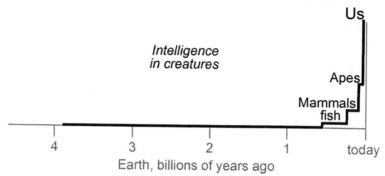

Earth, billions of years ago

For billions of years, nothing much happens. Then there are creatures with simple brains, then animals with backbones and a little bit of intelligence then, about 250 million years ago, mammals, then apes and porpoises, quite intelligent, and then us, very intelligent. The process driving evolution suddenly becomes super-efficient at making more intelligent creatures.

How efficient? The evolution of human intelligence involved a tripling of the size of the brain and the creation of our huge forebrain in just 5 million years.

Looking at the evolution of intelligence suggests this: it accelerates even faster than evolution as a whole.

Note how precisely opposite this rise of intelligence is to the chart of the creative capacity of natural selection.

What does this diagram suggest? To me it suggests that the evolution of intelligence is a runaway process. The faster it goes the faster it grows.

We've come up with two attributes of the mechanism behind evolution. It gets more efficient as evolution proceeds. And once the evolution of intelligence gets under way it goes even faster, it seems to catalyze itself.

Now let's see if we can figure out where such a mechanism could be located.

CHAPTER 3

Candidate for the role of "Violent Engine"

What can you think of that started out at the same time as evolution, accompanied it all along, getting more and more elaborate all the time? All I can think of that fits this description is the genome. It's been passed on intact, without a break, generation to generation, from the beginning, getting more elaborate with time.

How capable is it? We all start off as a single cell; from that cell grows a new human being, complete with its brain elaborately configured and wired. Where do the specifications for all that wiring lie? Where else but in the chromosome? Brains don't give rise to other brains, they rot with the rest of the body, all that gets reproduced, precisely, from one generation to the next, is the genome. If it can make us and our brains, it's plenty capable.

So when I look for the material support of whatever it is that drives evolution, I can't see anything else in the universe that fits the bill better than the genome.

Let's label the genome as the source of intelligence, and add that to the last diagram (see next page).

To me this looks like something acting intelligently. It creates intelligence in living creatures, and gets better at doing that very fast, as if it's learning how.

Could the genome support an intelligence, the way our

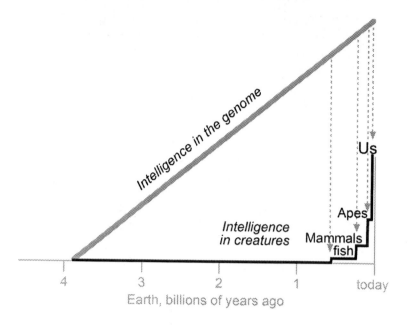

brains do? Are there processes in the genome that could make it act like a brain? There are. Examples are the exons and introns, alternative splice genes, methylation and other chromatin marks, small interfering RNAs that can migrate through the body, and protein-folding control. And scientists have only just begun discovering these mechanisms. The human genome is written in 3 billion characters, half as many as there are people living on Earth. If you strung those characters like beads on a necklace 8 to an inch it would stretch across the Pacific Ocean with 1000 miles spare. And it's been evolving 10 times as long as brains have, without a break. You can't logically set any limit to the machinery for supporting intelligence in the genome.

With that much intelligence to draw on, us getting our intelligence in only 1000 times as long ago as the pyramids got built seems more plausible—our intelligence didn't have to be created from scratch, it only had to be transferred to our brains from a source of intelligence that already existed.

We've got our first glimpse of a new candidate for what drives evolution.

What's it like? Not anything like a god. The genome isn't eternal or omnipotent. It's only as old as the Earth or less, and it screwed up big-time, over and over. It creates, it learns, it plans, it builds on past experience, it makes mistakes. What it looks more like is, one of us. Which is just as well: what we're looking for, after all, is the source of our own selves.

CHAPTER 4

Source of selfs

Were you surprised where we ended up in the last chapter? You shouldn't have been. Ask yourself: "What must the world be like if it can contain conscious decision-making selves like ours?" One obvious answer is, it can contain other conscious decision-making selves besides our own. If our self can be creative and drive chemistry through conscious decisions, so probably can other selves in the world. That's what a self-based science would start out assuming!

What in the genome would correspond to brain chemistry in us? Let's suppose, the chemical units making up the chromosome—all those "beads" I mentioned—that spell out our genes, plus all those "exons and introns, alternative splice genes, methylation and other chromatin marks, and small interfering RNAs."

In us, thinking and our brain chemistry can drive each other. Brain chemistry can shape some of our thoughts, then those thoughts lead on to other thoughts that can act back on brain chemistry. Maybe the same thing happens in the genome: its thinking and its chemistry can drive each other. A "pool" of genes can generate some of its thoughts, which lead to other thoughts which then act back on those genes to change them.

So far I've been following a fairly logical path. After

coming up with a new set of assumptions I looked at the whole path of evolution to identify the most likely location of what drove evolution. That pointed to the genome itself. Because our science has been self-based, it wasn't a great stretch to imagine the genome having a creative decision-making self attached to it.

But I think you're entitled to be surprised by what comes next.

Origin of variation

What will happen as the genome thinks? Just as we can change our brain chemistry by thinking, the genome's thinking will change its chemistry—the genes on its chromosomes that code for living creatures. You could say, just as we "dream up" new designs and stories, the genome "dreams up" new creatures. In my new self-based science, this is the source of variation. The genome thinks variation into existence.

But what leads to adaptation? How can new variation dreamed up by the genome result in creatures getting better adapted to their surroundings?

Just as it's possible to come up with ways natural selection and mutation can lead to adaptation, it's possible to propose ways the genome-self dreaming up variation can lead to adaptation.

Let's ourselves "dream up" a new origin story.

A new origin story

Chapter 1: Think back to when the infant Earth was about half a billion years old. Simple physical mechanisms created the first proto-genome. With the birth of that proto-genome, a tiny intelligence popped into existence.

All that tiny intelligence knows is, as it thinks certain

kinds of thought it grows. What's happening is, its thoughts cause changes in the genome, which of course code for living creatures. When those living creatures succeed in reproducing and making new creatures like themselves, the result is new copies of the genome that reinforce the genome's self—it grows. By thinking like this, being "reinforced" by thoughts that make it grow, it learns to make new species of creatures better and better adapted to their surroundings and so more likely to survive and reproduce.

Chapter 2. As the genome-self learns what kinds of thoughts lead to creatures that are better adapted to their surroundings, it starts to reconstruct, or "visualize," the physical world those creatures inhabit. Now it can think thoughts about this world. It can even think up "scouts" with new talents and senses better equipped to tell it more about this world. Through these "scouts" it can imagine the physical world more and more clearly.

Chapter 3: After four billion years, the genome-self starts thinking about itself. As it thinks about its self, its thoughts code for new genomes for living creatures with brains, into which get embedded some of its own intelligence, along with a bit of its own self. The most recent of these thoughts is us.

End of story.

Our origin

First, by imagining a genome self able to think as we do, we arrived at a new origin story—the genome-self thinks new variation and new living creatures into existence. Now, what kind of story about ourselves can we turn that into?

We appear as the most advanced of this self's scouts. To make us better scouts it built marvelous new talents into us, ready to be drawn on and developed by civilization. There may be many more talents in us than we've yet discovered. What skills does it take to run a cell, to grow a human body, to repair a lost limb, to design a new creature? All that science, all that mathematics, all that art, may be within us, waiting to be discovered. All we need is a reason for believing they're there. The potential for self-enhancement from such a self-based science has no bounds.

A better explanation?

Any new theory like this seems far-fetched at first. Is it worth taking seriously? That depends on how many puzzles it solves.

What puzzles evolutionists most today is not the origin of species so much as the sudden appearance of major new divisions in nature, such as creatures with backbones and animals with brains. Curiously, these major steps sometimes seem to occur before they're actually needed, they're "pre-adaptations." All this is easier to account for with self-based science. The genome-self thinks up new kinds of creatures to explore new territories.

How about the puzzle of our own origins? Instead of thinking of ourselves as apes descended from an ape we could say we are selves descended from the genome's own self. The ape-part of us would simply be the platform the genome-self used to embed a part of its own self in.

In this new science, what more significant is not what separates different kinds of creatures, but what connects them. For example, what connects us to our pets?

One of my cats is black, with green eyes. When she wants me to open a door for her she comes and fixes me with those bright orbs. When she sees she's got my attention, she turns and darts off a little way, then turns back to see if I'm watching. Then she darts off again. Once she's drawn me to the door, she gazes up at the door handle. Clearly she knows she's something that can be seen. She also knows that she can change how she looks to tug on my altruistic impulses. In other words, she shows signs of knowing that she and I both have selves, and that we can expect to understand each other. Incidentally, she is also exhibiting tool-making behavior. Cats did not evolve to get other creatures to open doors for them. They've figured out how to use various parts of their bodies to make tools out of us.

Now think about this: cats separated from the line leading to humans about as long ago as when dinosaurs walked the Earth. We're very distant cousins. Most of the things cats do seem utterly alien to me. Yet when we play there's a genuine meeting of selves. If there wasn't we wouldn't do it. Despite us being two very different "platforms," self-based science suggests we're connected by overlapping portions of the same self having been embedded in us.

And in a garden I see design and form everywhere, much more form than I'd expect adaptation alone to induce. The riot of form in a garden or a hedgerow looks like the work of a passionate intelligent artist, the whole landscape seems charged with intelligence.

Self-based science starts out just the opposite of today's science. At the center is the conscious human self, and from there it works outwards in wider and wider circles until its outer boundary meets and merges with today's material science. Replacing natural selection, it makes science complete. It can provide us with explanations for parts of our experience that matter very much to us, about which natural selection remains silent.

For more speculations turn to "Self-based science" in the "Fine Print" Appendix, starting on page 123.

Natural selection is no longer proof of Physicalism

We've played the science game. We've set up assumptions, we've drawn conclusions, we've identified a possible new mechanism for evolution. There are now two mechanisms proposed for what drives evolution, only one of which—natural selection—supports Physicalism. Advocates of Physicalism can no longer claim that conscious decision-making and the self are ruled out by evolutionary science.

At last we're equipped to venture out into the world, to challenge natural selection and strip Physicalism of its support.

Part 3

Manual:
How to save the self

Advanced study
Dream on
Are we a pre-adaptation?
What should we tell the children?

CHAPTER 1

Threat of a new barbarism

Just as you don't shout "Fire" in a crowded theater even if there really is danger of a fire, perhaps you shouldn't teach Physicalism in the classroom even if you think it's true. Your action may just make things worse.

How can Physicalism be a danger? Because the authority of science carries it into every corner of our culture, in every country. This threatens us with the awful possibility of a worldwide slide into barbarism.

Unlikely as it seems today, civilization can plunge rapidly into barbarism. It happened 4000 years ago to the great civilizations in Sumeria, in what is now the Middle East. What was then a civilized world plunged into a thousand-year Dark Ages of brutality and ignorance. Something similar happened when the Roman Empire in the West collapsed. It took only a couple of generations for people to become brutal and illiterate. Except for isolated and defensible cities like Venice, civilization survived mainly in outposts of the old empire such as England and Ireland. The core of the empire, in just a few generations, reverted to barbarism.

Barbarism is still found occasionally in the form of "feral children"—people who go through childhood without human contact and without learning any human language. From Wikipedia: "they cannot show empathy with others.... are usually entirely unaware of the needs and desires of others. The concepts of morals, property and possessions are alien to them." Complete social collapse could lead to something like that, to the loss of everything we regard as distinctively human. There is no limit to the potential loss of human nature when you plumb the depths of barbarism.

The new barbarism

Distressingly common is a descent into partial barbarism, such as "showing no empathy for other people." Perhaps most curious, though, is people who take the extreme position of showing no empathy for their own selves.

Here pride of place goes to Victorians who embraced a curious doctrine that denied them a self of their own. Their own self they called an "epi-phenomenon," something with the appearance of being real but not having any reality in practice, not capable of having any effect on the physical world. It's Physicalism as experienced from the inside. "My conscious self is not what directs my behavior in the real world." I even subscribed to this bizarre belief myself for a few years until the contradictions built into it became too obvious to ignore.

Physicalism and epiphenomenalism may be the most direct route to barbarism. Within a few generations children brought up by parents with those beliefs would have no sense they could shape their selves by deliberately learning anything new or by applying reason consciously. Whatever they did they'd have no thought of questioning it. This would be barbarism by deliberate adoption. And if such a culture was adopted worldwide there would be no one left to borrow civilization from later.

By itself, Physicalism is harmless. Everyday conscious

experience contradicts it. It becomes dangerous only when the enormous authority of science is placed behind natural selection and how it denies consciousness.

Here are ways natural selection is already preparing the ground for a new barbarism.

The new name—"Naturalism"

Just as academics don't wear their cap and gown when they go shopping in the mall, they don't call themselves "Physicalists" outside the college campus. Instead they call themselves by the warm fuzzy name "Naturalists." Members of one growing community of Physicalists, with its sights set on political action, have adopted the name "Brights." From their web site (www.the-brights.net): "The ethics and actions of a bright are based on a naturalistic worldview....join with other brights from all over the world in an extraordinary effort to change the thinking of society...We want persons who hold naturalistic outlooks to be elected to public service and to be able to participate openly and candidly in civic affairs." Prominent among "Friends of the Brights" listed on the site are Richard Dawkins, Daniel Dennett, and Steven Pinker, Physicalists all.

The new eugenics

First time around, eugenics was coercive. If some people were judged to have genes the rest of us didn't want in the gene pool, they got sterilized.

The new eugenics is voluntary. Women can apply to a sperm bank for semen from men with just the characteristics they admire and want in their children.

See the difference this could make? If these women want their children to succeed in the face of natural selection, they'll select for high levels of testosterone, physical strength, aggressiveness, callousness to anyone but family members, and promiscuity. On the other hand, guided by a self-based science they'd be more likely to select for intelligence, sensitivity to others both human and otherwise, design and creative ability, and talent for languages.

We're no longer dealing in theory. We're talking about acceptance of natural selection causing rapid genetic changes in human nature. Which science we subscribe to becomes a potent force in the world.

The new economics

Economics got called the "dismal science," I suppose, because it seemed to have so little to do with human nature. But that's far from the truth. There's always a theory of human nature behind economics. That's the point. It's for turning theories of human nature into industrial plant and consumer products, into the goods we surround our selves with.

The human talent for bargaining and making deals is the basis for a modern market economy and the stock market. Economists now find that idea naïve. People really aren't good judges of what's to their best advantage. They behave irrationally. So economists are beginning to turn to natural selection as a guide to what human beings are "really" like. If we don't watch out, that's what our world will increasingly come to reflect.

Imagine, though, economic theory built around a self-based science. Then it would center on creativity, intelligence, learning, and mutual interest. The difference would literally reshape our world.

The new literary criticism

There's a new literary criticism based on natural selection. We're already familiar with its main ingredients. From cheap fiction it borrows the motivations of sex, power, plus a little nepotism. From Wall Street it borrows the emotions of fear and greed. To reveal what the classics are "really" about, practitioners of the new literary criticism rewrite them—Jane Austin's *Pride and Prejudice* for example—in terms of just these motivations and emotions.

The new education

To help our young people triumph in the natural selection stakes, shouldn't we train them in aggression and promiscuity? If you accept natural selection as our origin story it's hard to quarrel with the logic of this policy, though you might not like the results.

The new psychology

In the form of "evolutionary psychology," natural selection is already casting dark shadows over human nature.

This new psychology has several strands. One is, how we adapted to our environment tens of thousands of years ago accounts for impulses in us today, such as our tendency to overeat when food is readily available.

Another strand uses game theory to account for how we arrive at our decisions. Hundreds of thousands of years of trial and error have supposedly taught us to use winning strategies in our relationships, such as "tit for tat"—trust first time, then trust or cheat as the other guy did last time. Either these winning strategies got built into our genes, or our genes have evolved the ability to figure these strategies out in a split second each time. Inheritance of these strategies makes consciousness redundant.

A third strand of this new psychology reminds you that the genes that end up in the gene pool come from individuals who, one way or another, managed to survive the process of natural selection and had most progeny, those progeny in turn going on to survive and reproduce. To be successful in the natural selection stakes as evolution demands, ask yourself, what would favor your genes being passed on? The answer from evolutionary psychology: aggression against competitors, promiscuous relations with the opposite sex, a total disregard for morals, and nepotism towards your children. Everything about us has to be explained in terms of the small set of grubby motivations and emotions natural selection can account for. It's not pretty.

Friendship? Is that important to you? Enjoy it while you can. To evolutionists it's their worst enemy—altruism to non kin—so it has to be explained away. First, it's identified as a strategy: "I'll help you now so you'll help me later." When humans first left the trees and began hunting in the grasslands, the story goes, this strategy contributed to our survival, so it got built into our genes. Those feelings you enjoy when you're with friends, they're just a by-product of this strategy. Ignore them—what matters is making sure you get back as much as you give. That's what friendship is "really" about, that's what those feelings are "for"!

Even those feelings have to explained away. In men, what looks like friendship is really "nothing more" than how strangers relate to each other. In women, friendship is really "nothing more" than treating strangers as if they were blood relatives. What we experience as the distinctive warmth of friendship is "really" made up of feelings that go along with the natural-selection trinity of sex, power and nepotism. And that goes for all our other warm fuzzies as well, such as compassion and concern for humanity in general. We're hanging on to traditional misconceptions of what our feelings "really" mean. Let go of them! Get over it.

The new self

This message of evolutionary psychology is spilling out into self-help books with titles like "Mean Genes" and magazine stories telling you how to really "know" yourself. The ground's been well prepared. Physicalists' insistence on teaching natural selection in the classroom is doing its work. Values in our culture are undergoing a massive shift. And any opposition to this shift is waved away as fanatical Creationism.

This is the battleground for the fight to halt the spread of Physicalists' primary agent, natural selection. Let's move on to tactics.

CHAPTER 2

Tactics for attacking natural selection

Strategy: "Lay off the self, or we kill natural selection"

Natural selection is protected by the greatest power in the land—a lot of jobs depend on it. Many of them are highly paid professional and university careers. So we're not going to defeat natural selection.

But people in those jobs often depend on government money, foundation grants and college endowments. They're very sensitive to criticism. So I recommend just trying to harass and embarrass them enough to bring the issue to their attention, to make them promise they mean no harm to the conscious self and that from now on they won't let Physicalism be taught as science in schools.

We want that guaranteed in writing. Maybe as a sticker to go on school biology books!

Speaking out

As soon as you speak out against natural selection, you're likely to attract one of Darwin's Bulldogs. You'll be told you're stupid, you're ignorant, you should read some books on evolution, you don't know what you're talking about, and so on. Two chapters in my novel *Me and The Genies* contain samples of this kind of abuse. It's not pleasant.

Walk away. These are usually ignorant young hotheads

totally unreachable except through the people they get their opinions from. Those are the people you need to talk to.

If you do find yourself talking to actual evolutionists, be aware that to them being asked to question natural selection is like being asked to question whether 2+2=4. That's how remote it is from their day-to-day professional life. You might as well be asking them to reach back to primary school and remember what you're supposed to do before you cross the road. So, instead, tell them why you care. "I'm upset when I hear about kids in school being taught as part of biology that they don't have conscious decision-making." Evolutionists will have heard that a million times too, they argued it out as freshmen in college, it's way over and done with, a closed book. OK, just let them know it's not a closed book for you. If they don't glaze over or brush you off, ask, "Doesn't just learning science involve conscious judgment? Can kids learn science and that they don't have consciousness, both at the same time?"

You can't win an argument about whether natural selection is true with people like this because they'll start talking about how mutation proves natural selection is true and so on. The most you can do is be part of a growing number of people expressing concern about the issue, about their children being taught that they don't have conscious decision-making, and is that something to do with natural selection?

The rise of Creationism has made evolutionists very sensitive to this kind of criticism and they'll want to place you along the natural selection-Creationism spectrum. To distance yourself from Creationism you might throw in the phrase, "Not because of religion, just because I believe our conscious decisions can make a difference."

Notice I use the phrase "conscious decision-making," not "consciousness." Consciousness is too woolly. But conscious decision-making is about the cause of things happening in the world, something scientists can deal with.

Don't be an anti-Darwin bulldog. Don't go on too long. Don't get involved in technical matters, especially mutation. Instead keep the discussion on a personal level.

A good topic to bring up is friendship. How can people who think they're robots make friends? Why would they? What does friendship mean to them? Is it just a "strategy"?

Other tactics: letters to the editor

When you see articles in magazines involving any of the "new" topics I listed in the previous chapter, that's a good opportunity to write a letter to the editor giving your point of view. Magazines don't get many letters so you've a good chance of getting published.

How about movies? For example, both the movie and book titled "Thank You For Smoking" were about someone making a career out of lying to the public for money. But they made entirely opposite points about it. A skilled public relations professional is shown "spinning" information—in this case about the effects of smoking—to imply just the opposite of what he knows the information actually means. You know this because, along with fellow PR flacks for the alcohol and firearms industries he refers to himself and them as "Merchants of Death." He even explains to his son some of the tricks he uses.

By the end of the book he's had a change of heart and uses his PR skills to oppose such "spinning." The movie instead has him becoming a successful consultant applying "spin" for new clients and his "Merchants of Death" club has expanded to include several new members. His son is shown basking in approval for winning a debating contest, presumably by using the tricks his father taught him, and appears destined to follow in his father's footsteps. The "natural selection" take-home message of the movie, exactly the opposite of the book, is: all that matters in life is success in the struggle for power, survival and sex.

What can you do when you see something like this? Tell people. Visit online blogs about the movie and point it out. Write to the movie-review editor of your local newspaper. Movies live or die by word of mouth—that's us, talking to one another about them.

Display videos of cell division at science fairs

If you get an opportunity to prepare an exhibit at a science fair or to help someone else come up with a subject, you could display a video of chromosomes during cell division. At the time I'm writing you can find some at www.bio.davidson.edu/Courses/movies.html—check out the merged videos of "mitosis" and "meiosis," and the non-merged videos of mitosis.

I vividly recall the first time I saw such a movie. I was dumbstruck. Here were dozens of individual molecules, I had to remind myself, bunching up big enough to be seen under the microscope, and engaging in a fantastic synchronized dance, to make life. I do recommend taking a look. It is one of the most amazing sights in all of science. It's almost impossible to watch this without an eerie sense of an intelligence at work. It certainly elevates nature far above commonplace notions of "chemical reactions."

Join debates on "teaching the controversy"

Up to now, we've let Creationists do our work for us, keeping natural selection out of the classroom. I say, it's time to step smartly in front, and take over. Yes, there is a controversy, but it's not over whether to include Creationism alongside natural selection in the classroom, it's about whether natural selection itself should be taught. Evolution, yes. But natural selection, no.

I encourage you, sign up to speak on the subject. And when you do, speak against both natural selection and Creationism being included in classroom science.

Once you speak about this in public you enter the arena known simply as "Origins." This is pure pitched battle. The two sides scream at each other, there's no pretense of listening, no limit is set to the abuse heaped on their opponents by either side, it's just brutal "us against them."

I suggest, say your piece, vote, and then keep quiet. You might say, "I want children to learn about evolution but I don't want them taught about either creationism or natural selection. I think natural selection implies we're not capable of conscious-decision making." If you're asked to identity your belief you could say you believe there's an intelligence in evolution like our intelligence.

Encourage new thinking in evolution

A self-based science has fantastic potential. Whatever we discover about what makes evolution so creative, we could apply to our own selves.

But I have to tell you, there are huge problems with it. I mention some of them in "Self-based science" in the "Fine Print" Appendix starting on page 123. Try explaining self-based science to people and they'll think you're crazy.

So don't try to persuade other people about the merits of self-based science. But whenever you come across scientists studying consciousness, applaud them. Write letters in support. Express your interest. Write to your legislators to say you want this research funded.

If we can generate sufficient influence, let's press for a high-level commission where the world's most eminent physicists, mathematicians and philosophers of science judge the soundness of evolutionists' claims and methods.

CHAPTER 3

Living well is the best argument

Can we turn natural selection against itself? Suppose we adopt a self-based lifestyle and it spreads and displaces the natural-selection lifestyle of "evolutionary psychology" and "Mean Genes." Wouldn't that prove that self-based lifestyles are "fitter" as well as better than lifestyles based on today's evolutionary psychology?

Let's give it a try. Let's sketch out a self-based lifestyle.

Hidden talents

We self-based folk, we know we've lots of talents embedded in us by the genome-self, we don't even know how many. A lot more than we'd get just from "adapting to the environment." Whatever we see in nature we can say, "I wonder if that got built into us as a talent?"

Balancing on two legs is quite a talent. It's very efficient—we're nature's long-distance running champions. More extraordinary though—and maybe what it's really for—it freed our forelegs to become arms and hands with even more remarkable new talents. This must have been important, it all got rushed through in just a couple of million years.

One of our "handy" talents is drawing. Our hands become another way of expressing ourselves. I can testify to that. In my forties I decided to learn how to draw. I set myself exercises and learned how. Now I can express what

I think as a drawing. Not only that, but I love doing it.

The real "talent" is not just drawing, it's the eye-hand loop. That's a feedback system that can be harnessed to do just about anything you can imagine, like drawing, making tools, playing musical instruments, knitting and other crafts, sketching building plans, making and using puppets, ping pong, driving, making and racing model airplanes.

Look around you and you might say, "I want to create the kind of beauty I see in nature." And, if you assumed that you had nature's talents built into you, you might just sit down and practice until you succeeded. You've a new talent. You wouldn't bother, though, if you assumed talents could come only from adaptation through the selection of mutations.

This world's sensational

Ready for some heightened conscious experience? Turn up all the instruments in the orchestra and prepare for a performance. Go to your front door (assuming you don't live in an apartment). Check off your senses—sight, hearing, sense of your weight on your feet, your body's posture, what you can see of your body out the corners of your eyes (very important), your surroundings and how they enclose you, the air temperature. Open the door, and walk out. Symphony!

Next, let's see how the senses perform when we take them completely out of the environment we evolved in, such as driving a car at 60 miles an hour on country roads while playing Tchaikovsky at full volume, which I like to do. Then, lead forward, rest your folded arms across the top of the steering wheel and close one eye—closing one eye suppresses three-dimensional vision, flattening vision to photographic scene-making. The car will drive itself as you lean in the direction you want, while you just enjoy the pictures emerging ahead of you.

We weren't exposed to any part of that 50,000 years ago, yet we can do them all at once and enjoy it. Account for that through evolutionary psychology? I don't think so!

You become what you pay attention to

If you live as if what matters most is passing on your genes—choosing a mate, having lots of children and giving them a good leg up in life—then all of life's big challenges will hit you between the ages of 15 and 35. For the next 60 years—from 35 to 95—you'll matter less and less, to yourself and others.

But if you live as if what matters most is your self, the more years that pass the more opportunities you get to enrich your conscious experience. The self is where we evaluate things and decide what to do, and that's what shapes most experience. Invest in the self and as time goes by you'll become more valuable to yourself and to others. You enrich your experience and theirs.

It's just a matter of figuring out what to invest in.

For me, part of that is setting up filters for what I don't want to let in. Some things I've learned from experience just don't interest me—romance and crime stories, for example. I filter them out. I've no competitive instinct so I screen out all sports. I've lost interest in art so I now I filter that out too. I pay attention to what I do respond to and what I don't respond to and I install filters to match.

I've learned that violence scramble my attention so I filter that out. When I heard about the Twin Towers being hit I installed a filter for that. Only a week latter did I let myself read a print account of it. I still haven't seen it on television. I'll keep that filter installed until the event becomes just history. Not because it isn't important, but because too much exposure to it would scramble my judgment. I think a lot of people had their judgment scrambled by it.

How about science? I enjoy news of science and technology. But I filter out news of brain function and most medicine because it reinforces how we're made of matter at the expense of the conscious self. I take sides even with science.

Don't worry, no matter how much you filter out there'll be a lot left. The media make sure to bombard us with so many messages there'll always be enough getting through.

Then, the attention can be primed. Going into New York City for a few days? Watch a video on the history of skyscrapers to make your experience more interesting. We're lucky to live at a time when there's so much "priming" to choose from. The capacity of our memory seems to be infinite, you can stuff it full every day and there's always room for more. The more you pack in, the richer your experience gets.

If you do this anyway, if you value the quality of your experience and do what you can to enhance it, you're already living a self-based life.

Be a good scout

What are we here for? According to my self-based science origin story, we're scouts the genome-self dreamed up to explore and report back about our environment. The genome-self equipped us with superb senses. Now what?

Use those senses. Be sensitive to them. Read philosophy, learn what questions people have asked about them, then come up with new ones. Conceive as wildly as you can. Then check things out. Relish. Mull. Develop talents. Look for new ones.

Are you a good scout? Here's the test. Find a public park. Sit on a bench alone for half an hour. Were you bored? Then you're not ready. Read some more books, study birds, talk to new people, and try again. Keep trying until you get it right.

Advanced study

Once you pass these tests, you're ready for advanced study. Get an animal companion, what you're probably used to thinking of as a "pet." Learn from each other.

Dream on

I said, we're scouts the genome-self dreamed up to explore and report back about the physical world. I've suggested ways to explore. But how do we report?

Perhaps through dreams. In my dreams I'm presented with elaborate scenes that I get caught up in. What's that about? It's as if I'm being put through my paces.

Maybe I am. Maybe the genome-self is querying me about what things mean. It's making me report back.

And not just me. When my cats sleep, I see their eyes flickering just as ours do when we dream. Animals have probably been dreaming for a couple of hundred million years. Perhaps they too are being made to relive their day and report back. Like me, when they wake they probably think back to the adventures they had in this or that dream.

Maybe that's where our sense of a self and consciousness comes from. We carry a conscious self from our dreams into the light of day. Who am I? I know who I am, I'm the self I experience being my dreams. The genome-self remakes my self each night during my dreams. Maybe consciousness first got invented not for daytime, but for nighttime.

OK, reality check! That's far-fetched, to say the least. But some dreams must come out of the genome. As a boy entering adolescence I experienced extremely erotic dreams showing me in vivid detail things I'd never seen before, let alone done. Though I've not asked around, I assume all men have had this experience. Where else, but from the genome, could these dreams have come, preparing me for my role in the reproduction of my kind? If that dream can come from the genome, why not all the rest? All those cave paintings from 30,000 years ago, what were they about? Reporting dreams, I believe.

Can natural selection account for our dreams? Could they be composed for us by mutations being selected? Time for another reality check.

Are we a pre-adaptation?

Biologists will be holding their sides laughing at me suggesting our dreams are laid on for us by the genome. But they do take seriously another very unlikely scenario, that creatures sometimes appear pre-adapted for some new role before they actually embark on it. Natural selection couldn't account for this, but of course, the genetic-self can, once it gets wind of a new environment it wants explored.

What about us? 50,000 years ago we already had all the wonderful new talents we've recently discovered in ourselves. But back then we weren't exploiting any of them. Were we pre-adapted for something? Civilization, perhaps? Science? Space travel?

I don't know. Just asking.

What should we tell the children?

Should we tell children they're nothing but a mix of mutations that got passed on during sex? That if they want their genes passed on, they should make sure they're the ones that survive and that they get plenty of sex. Is that the message?

Even if it's true, should they all be taught that message?

And if it's not true, should that message be taught, to anyone?

Some advocates of natural selection mock me by saying, "So according to you we should just vote for what's the truth, as if it was a popularity contest!" That's as opposed to what they think, which is that science is beyond criticism and tells the truth no matter what the rest of us think. I think the history of science says otherwise. The one-in-four of academics that don't experience decision-making in consciousness are writing all the books. I think it's time for the rest of us to have our say. Yes, let's vote for the truth.

I say, truth is too important to be left to the scientists.

APPENDIX

The fine print about…

THE FINE PRINT ABOUT...
Physicalism and brain chemistry
See Part 1, Chapter 1

Physicalism

Take an idea, like "democracy." Does it exist? It may—Physicalism doesn't necessarily take a stand either way. All Physicalism says is, only through physical processes can it make any difference. How about the self? How about conscious decision-making? How about what we used to call "mind"? For all these mental entities, the answer's the same. Physicalism doesn't make claims about whether or not they exist, only that to make a difference in the world they have to operate through physical processes.

And logically you can't argue with that. If you define "physical processes" as "every possible cause there is that can make a difference" then by definition Physicalism is true, it becomes a tautology.

But what matters is how the phrase "physical processes" is used in practice, and in most cases it's used to mean that the only processes active in the world are the physical processes *we already know about*. All those entities I mentioned above, they all may exist, but the only way any of them can make a difference in the world is through physical processes we know about *today*. And since it seems you can't have mind or consciousness without first having a brain, the processes all those mental entities operate through must lie in brain chemistry *as we know it today*. So, according to the Physicalists, only through brain chemistry—as we know it *today*, since we understand matter *completely*—can consciousness have any impact on the world.

In arguing about physicalism, it's good to keep people honest. Are they referring to physical processes *as we know them today*, or in some hypothetical future when our knowledge really is complete? If they insist the science of matter is complete *today* you can politely agree to disagree.

Brain chemistry

Again, if you define brain chemistry as able to support absolutely everything about "consciousness" and "mind," you can't argue with that either. But once again it's how the idea gets applied that matters. And how Physicalists define brain chemistry is, chemistry as we know it *today*. Like the Krebs Cycle. That's a key biochemical pathway in the body with the same chemical passing through a series of chemical reactions and releasing energy as it goes. Each chemical reaction is the result of the one before, and the starting point of the one to come. The connections between them are straightforward chemical reactions. And that's what scientists mean when they say consciousness exists only as chemistry. In other words, as in the Krebs Cycle, each step in conscious thinking must be connected to the next by a simple chemical reaction.

Of course, it's not so simple to decide whether conscious decision-making actually is or is not just brain chemistry. The crucial question is, *today*, are you better off thinking of conscious decision-making as being chemistry, as we understand it *today*, or as being a succession of non-chemical thoughts that can drive brain chemistry?

You could spend your life trying to figure out how the chemistry in the brain works, but what's the point? You already know how the brain works—it does whatever it has to to make conscious decision-making happen. There

probably isn't one particular chemical equation that accounts for why a scientist makes one scientific judgment
instead of another. It'll happen through one reaction one
day, through another reaction the next day. The chemistry's probably always changing. What stays the same is the
result—a sequence of conscious thoughts linked to one another not by chemistry but by what each thought means
and how that meaning connects it to the thought that follows.

A little help from my friends

Within my local humanist society I've two Physicalist
friends I can bounce ideas off. With their permission I'll
share their responses to a remark I posted about the term
"free will." I wrote:

> I've switched from talking of "free-will" to "conscious deci-
> sion-making," and shifted the focus of attention to a new
> situation: the practice of science. When a scientist declares
> something new to be scientifically valid (eg in peer review),
> does that involve conscious judgment? The interesting
> case is, if it doesn't, what is the validity of that declaration?
> Is that declaration no different from, say, one step in the
> Krebs Cycle, just a chemical reaction driven by other chemi-
> cal reactions? Or do we assume there's a non-Krebs-Cycle-
> type connection between phenomena and conclusions about
> scientific validity? Is judgment about scientific validity a
> predictable formulaic process? How would you characterize
> the critical activity in making such a declaration, where I
> would call it conscious decision-making or judgment?

Here's Mike's response:

> Folk psychology tells us that we have a conscious mind and/
> or will that causes our bodies to do things. Two problems

- nobody has ever been able to find evidence for or investigate the conscious mind other than one's own, and BTW, how does this mind cause our physical bodies to do things? What interaction is involved? Nevertheless, there is a strong subjective feeling that we have conscious minds that we use to control our behavior.

Get used to your experience of conscious decision-making being referred to as "folk psychology." Then Mike brought up the Physicalist claim that only physical causes can have physical effects.

The two problems above lead nearly all investigators to discard dualism, the idea that minds and bodies are different things. Instead, the claim is that the mind exists and is instantiated in the brain and body. This is a materialist view but does not say that mind does not exist.

Physicalism doesn't deny mind exists, Mike agrees, but he does insists it's "instantiated" in matter, which I take to mean it's a passive reflection of what's happening in the material that supports it. But:

One objection to materialism is that it seems to deny us free will. So we the a dilemma - either some sort of dualism is true, which means we will never get a handle on psychology; or materialism is true, and we have no free will.

A way out of this dilemma is offered by analyzing the concept of free will to see if it is actually compatible with materialism. Daniel Dennett has done so. His conclusion, briefly, is that we do have the varieties of free will that are actually important to us as humans, and that they are instantiated in our social lives, cultures, and brain processes. No, we do not have a radical kind of free will in which we can do things regardless of the material states or our brains and bodies. However, our minds, as instantiated in our brains and bodies, have evolved to be complex and intelligent so that we can weigh possibilities and base decisions on judgment, possible outcomes for ourselves and others, etc. Some aspects of the material processes constituting this decision-making are experienced as conscious, some are not.

See Dennett's *Freedom Evolves* and Wegner *The Illusion of Conscious Will.*

Dennett's *Freedom Evolves* is a curious book. It's very simple and clear, except for one section that's almost totally obscure. Unfortunately all the rest of the book refers to conclusions asserted in that one section. What Dennett seems to say there is, since measurements must have a finite number of decimal places, the future cannot be accurately predicted. Since matter's future cannot be accurately predicted, neither can our brain chemistry, a particularly complex example of matter. Therefore we are, to the extent of our inability to measure perfectly accurately, not determined. Our inevitable errors in analyzing brain chemistry provide the freedom of the will.

Satisfied? Could an inaccuracy in a scientific measurement account for your experience of consciousness? To me that's like saying an inaccuracy in measuring a lump of tofu can turn it into your left foot. I don't think so! The experience of consciousness isn't the same sort of a thing as a rounding error in a scientific measurement.

Here's the response from my other Physicalist correspondent, another Mike.

When a scientist declares something new, then yes, that involved a conscious judgment. And yes, that declaration is also the result of a series of electro-chemical reactions taking place within the matter physically making up their brain. Nothing more, nothing less. That's what ultimately underlies conscious. There's no "non-Krebs-Cycle-type phenomena", assuming you're using that phrase "Krebs-Cycle-type phenomena" in a very general way to cover all electro-chemical reactions that can take place in non-living non-conscious matter.

So yes, science does involve conscious decision-making, but no it involves nothing more than regular chemical reactions like those in the Krebs Cycle. Making scientific judgments about the Krebs Cycle would be just more reactions of the same kind. Are you reassured about the soundness of their judgments about consciousness?

A scientist making a judgment about scientific validity is in principle a predictable formulaic process, though in practice the brain is a very chaotic system with many complex feedback loops that make such predictions impractical. People generally have an abstract simplified mental model of other human minds that includes the concepts of consciousness and free will divorced from any underlying physical processes. Like all models, it is not a strictly accurate representation of reality, but it is something that people generally find useful for predicting how others will act in various situations. Sometimes this model is good enough to be useful, sometimes it is not.

Note, for Mike consciousness is only a model we apply to other people, not something we experience ourselves that we might feel the need to account for, that we might want to know more about.

Here again is the claim that, even though chemistry is deterministic, once it becomes too complex to understand it thereby acquires the capability of generating conscious experience. Note that this is an unfalsifiable assertion, since it makes a claim about something, consciousness, that it defines as impossible to verify. It is a leap of faith.

An earlier version of this was to suppose natural selection somehow harnessed quantum uncertainty to generate freewill. Now it's complexity theory. Same leap. Same faith.

I think we're all agreed that our understanding of chemistry is far too primitive to have much of value to tell us. Even if conscious decision-making "really" happens in the brain, *today's* science won't help us understand it, won't tell you anything more than what you know already from the inside—what conscious decision-making is like and how it works.

Conclusion: our conscious experience is the best guide we have *today* to how we arrive at decisions. And it probably *always* will be.

THE FINE PRINT
Science at a crossroads
See Part 1, Chapter 2

There's good news about science today, and there's bad news.

The good news

The good news is how little science has outgrown its roots in alchemy. Alchemy came to Europe from the East as a quest for the secrets of matter, especially the properties of the elements and the different species of metals, with success measured by an adept's ability to turn one specie or kind of metal into another.

Newton was an alchemist. Francis Bacon claimed he'd moved on beyond alchemy but all he really did was extend it to include physics and come up with better experimental methods and standards of proof. Positivism was very explicitly based on Francis Bacon's work, making the procedures much more rigorous and clearly defining what those procedures could and should not be applied to.

When alchemy first came to Europe it got caught up in Millennialism and the search for Adam's lost wisdom, recovery of which would bring on the Apocalypse and salvation for all Christians. But already with Francis Bacon that was changing and the goal became the application of material discovery to technology, to improving Man's lot not only in Heaven but here on Earth as well. And in Positivism this second goal became "Progress"—the unending improvement of human life through material technology.

No prophet in the history of the world ever foretold the future so accurately and had so much to do with bringing that future about as Francis Bacon. In just 400 years we've gone from crude clockwork, horse-powered mills and hand operated printing presses to nanotechnology and the mass manufacture of computer chips. In the *New Organon* Bacon focuses specifically on heat, an astonishing choice at the time when there were no ways of studying it. And in his Utopian fantasy, *A New Atlantis*, his ideal scientific society has a department of dealing with engines! Right there, by 1625, Bacon has pointed ahead to the steam engine and the internal combustion engine, keys to the new world he was bringing into being. 200 years later that success could be generalized into Positivism's perfectly reasonable expectation of endless scientific and technological progress that continues to this day.

This is all very very good news indeed. And a magnificent tribute to human ingenuity and talent, perhaps the greatest ever.

The bad news

The bad news comes from people trying to apply science beyond the limits wise men set for it. Even as Comte and his chief interpreter John Stuart Mill were drawing up the boundaries within which Positivism should operate, Victorian scientists were applying its powerful procedures outside those boundaries to species for which those procedures were never intended. The Victorians began looking for way to apply physics and chemistry to the transmutation of living species, to how one living species got turned into another. With the characteristics of one species in their alembic, they thought that by adding to it the potent reagent "mutation" they could use natural selection to condense the characteristics given off into aqua vita—spirit of life. Repeating this distillation in each generation, they believed, would result in the accumulation of characteristics needed for a new and more advanced species. Today's chemical analysis of the genome is the pursuit of that same alchemical goal—the transmutation of species.

The problem is, when you apply these procedures outside the material realm, the standards of proof developed over centuries of alchemy no longer apply. You can't do experiments and even if you could you'd have no way of judging the results. You've no idea what kind of logic to apply or what else to measure besides chemistry and physics. You can put a mouse in a flask and heat it until it burns to ash while collecting all the gases given off and compare the weight of the mouse before and after, but you've no idea if the ash and the gases account for everything that's been lost in the transmutation.

What you can do is, you can just assume that alchemy applies to the mouse and to living species and just assume that all that's involved in their transmutation is characteristics being reacted on by mutation and condensed by natural selection into new species. You can just assume that, and claim that Positivism "proves" you're right. To reinforce your Positivist credentials you can turn your theories into abstruse statistics. And this is exactly what happened.

Once again, the story revolves around poor Charles Darwin. Intoxicated as a youth with Positivism he applied its methods to the problem of the transmutation of living species—he even called his secret notebooks his "Transmutation notebooks"— and came up with a purely physical

and chemical reaction capable of achieving it, "Natural selection." But he realized he couldn't try it out, the evidence
for it was circumstantial, and he sensed there was a lot it
couldn't account for. He asked close friends for their opinions but they were uncertain. No one had ever applied this
way of thinking outside the chemical laboratory. Only 20
years later, when the young whippersnapper Albert Wallace came up with the same theory did Darwin abandon
his scruples and publish, though his book was a welter of
"maybes" and talked of several other mechanisms of evolution as well. The greatest scientific minds of the time with-
held their approval. Later Wallace changed his mind when
he found evidence that natural selection could not be the
whole story. But by this time it was too late. People with a
purely political agenda and no interest in the limits of alchemy harnessed the theory of the transmutation of living
species to their attack on the Church of England, and the
issue ceased to be, is this a good theory, to, are you with us
or against us? And that kind of a Shibboleth natural selection remains to this day.

Karl Popper added to Positivism the rider that for a theory to be regarded as scientific it must, in principle at least,
be falsifiable. Natural selection, he concluded, is not falsifiable and hence is not part of science. Elsewhere (page
107) I quote the evolutionist John Maynard Smith admit-
ting, "It is rarely possible in evolutionary theory to think of
a single decisive experiment or observation that will settle
a controversy." When challenging evolutionists on any issue to do with natural selection, consider asking, "Is your
position testable? Is it falsifiable? Is it possible to conceive
of conditions under which it could be proved untrue?" All

scientists are familiar with this condition of falsifiability and they will understand what you are asking them.

Science at the crossroads

There can be no question of science abandoning its spectacularly successful alchemical tradition and the technological progress that goes along with it. But what is science to do about phenomena that lie outside that tradition? What is it to do about the excursions it's already taken outside that limit?

It should retract the claims it's made in the course of those excursions, such as that the theory of natural selection has achieved the status of a confirmed truth. It should define more clearly what its methods apply to and what its tools can measure, and what they don't and can't. That would undo some of the damage this particular unwarranted excursion has done.

And though it doesn't have to do more than that, it might consider moving beyond the alchemical tradition and developing new methods and new tools, and new standards of proof, or even some other criteria for its explorations than truth. It could for example set itself up to be a teller of stories, the very best stories that have ever been told, that mankind will evermore tell and thank us for.

THE FINE PRINT
Why evolutionary statistics don't work

See Part 1, Chapter 5

Every so often insiders spill the beans about evolutionary statistics.

In 1944 appeared an extremely influential book—for example it inspired both discoverers of the chromosome's double helix—*What is Life?* by the physicist Erwin Schrodinger. How many factors does he say natural selection can distinguish between?

> It is essential that they be introduced one at a time, while all the other parts of the mechanism are kept constant.

Jump forward twenty years to 1967. At a picnic organized by another physicist, Victor Weisskopf, the mathematicians were stunned by the optimism of the evolutionists about what could be achieved by chance. So wide was the rift that they decided to organize a conference, which was called "Mathematical Challenges to the Neo-Darwinian Theory of Evolution." Nobel prizewinner Sir Peter Medawar opened it by saying

> The immediate cause of this conference is a pretty widespread sense of dissatisfaction about what has come to be thought of as the accepted evolutionary theory in the English-speaking world, the so-called neo-Darwinian theory.

One speaker "proceeded to show that if it required a mere 6 mutations to bring about an adaptive change, this would occur by chance only once in a billion years.

Twenty more years on, hearken to the eminent evolutionist John Maynard Smith in a 1989 essay titled "The Limitations of Evolutionary Theory." He admits mutation can't be measured.

> It is possible to measure mutation rates in very special circumstances in some microorganisms.... But in most situations, mutation rates cannot be measured...

It not just mutation evolutionists can't measure. They can't measure "selection" or migration between populations ("genetic drift"). Of these Smith says:

> Thus we have three processes which we believe to determine the course of evolution, and we have a mathematical theory which tells us that these processes can produce their effects at levels we cannot usually hope to measure directly. It is as if we had a theory of electromagnetism but no means of measuring current or magnetic force.... It means that we can think up a number of possible evolutionary mechanisms, but find it difficult to decide on the relative importance of the different mechanisms we have conceived...

"If the mutation rate is doubled, does this double the maximum rate at which a species can evolve?" The question has no simple and agreed answer, he says.

> But my reason for raising the question....was to make the point that a theory of evolution which cannot predict the effect of doubling of one of the major parameters of the process leaves something to be desired.

Summing up he says,

> It is rarely possible in evolutionary theory to think of a single decisive experiment or observation that will settle a controversy.

And jumping another two decades to 2006, similar warnings dot a recent textbook with the impressive title of *Making Sense of Evolution: The Conceptual Foundations of Evolutionary Biology.*

> Since drift is better thought of as the expected statistical scatter around the mean predictive fitness, it is a conceptual mistake to attempt to distinguish it from selection. It is, therefore, no wonder that the formal tools developed to do so are inadequate to the task...We should expect that it will often be impossible to detect selection in a straightforward manner through statistical analysis, and often impossible to show, even to a relatively high degree of confidence, that selection is taking place... (p 33); While natural selection may be the most important concept in evolutionary biology, empirical research designed to measure it in natural populations is (perhaps surprisingly) a relatively recent phenomenon....the standard approach...in fact falls far short of that goal (p. 36); We do not have (and probably cannot have) a theory for how G changes during long-term evolution... (p. 109).

So we've experts enough to cast doubt on the validity of the statistics that evolutionists depend on for their defense of natural selection and its various epicycles.

The basic challenge they can't meet is this: keeping a living creature so far from equilibrium with its surroundings (death) means keeping it in a dynamic state. Two things have to be balancing each other at the same time: the crushing tendency to return to equilibrium and become that pile of bones in a puddle. And, counteracting that, the "Violent Engine"—whatever really is driving evolution. Evolution is the shifting of that dynamic balance both back towards equilibrium (fish in dark caves losing their eyesight, whales losing their legs) and further away from it (our brains tripling in size in the past 5 million years). It's a balance between the automatic tendency to disorder, and something that's creating even more order (the Violent Engine). Out of that dynamic balance comes evolution. You can't say, with Fisher, we'll use natural selection only to account for the creation of new order. The greater problem is how to maintain the order you've already got. If you've already got something capable of the more difficult job of maintaining that old order it'll be quite capable of creating new order too. You no longer need natural selection. The Violent Engine can handle everything.

A little statistics of my own

Like a sieve, natural selection can make only one distinction at a time. Just as a sieve discards smaller pieces while retaining larger ones, natural selection discards creatures lacking any one essential factor, while letting only creatures with that factor through to reproduce and pass it on to their descendents. Let's call what a sieve does in one pass 1 unit of selecting power. Then per generation natural selection has that same 1 unit of selecting power.

Now, suppose there are 1 million factors (at least!) you want to keep from getting lost from "disuse," while losing those you don't use any more. Factors like that can get lost in around 1000 generations, so you need to keep selecting each factor you want to maintain at a rate of at least one thousandth of a unit-of-selection (.001) per generation. To maintain a million factors you need 1,000,000 x .001, or 1000 units of selecting power. But natural selection can generate only 1 unit. It doesn't provide you with enough selecting power to prevent a million factors from getting lost through disuse in 1000 generations.

Fisher ignored all that. He looked only at what you needed to select new factors. He said, take one factor needed for
some new feature. Even if it's only .001% more fit than the
alternatives, in thousands of generations that tiny increase
in fitness will keep getting selected and eventually sweep
across the whole population. Well, of course! That's like
compound interest. But "disuse" (the tendency of everything to revert to equilibrium with the environment) also
compounds over each generation, and easily swamps any
benefit from that one new factor Fisher's talking about.

All of modern population genetics is built on Fisher's
misunderstanding. So it all fails. You'd think evolutionists
would have found this out. But they've a problem. Their
math doesn't work well enough to show them their science
has gone wrong.

THE FINE PRINT
Disposing of more "epicycles"
See Part 1, Chapter 6

The epicycles
Competition

"Kin selection"

Why selection isn't important for evolution

Who's clustering the genes?

The origin of species, still a mystery

Wallace, co-discover of natural selection, turns against it

I'm going to reveal my secret, where I get my ideas from. I get them from making up stories. Like this—a story exploring the roles of mutation and natural selection.

Smoke and Mirrors
"Evolution Explained" the flyer said, and in Robert's handwriting "Hope you can make it."

He introduced me to the other speakers, then he pointed up. Far overhead ran a torrent of silvery fish, like a river without a riverbed seen from below. Above the river dark clouds roiled and lightning flashed. From out the bottom of the "river" fell a steady rain of dying fish.

"What keeps it up there?" I asked Robert. "That's what they're going to explain," he said.

Their answer consisted of lightning striking and injuring some of the fish, and the less-fit fish falling out the bottom. "...and that's what keeps the river of fish continually rising," the speaker concluded. "Way over there," and he pointed upstream, "it starts out at ground level, but by the time it gets here it's high above the ground, and rising even higher."

"I've a question," I said. "A few fish getting injured by lightning and the fish all fighting each other may be what gets them off the ground in the first place, but what's holding them up now, way up there?" and I pointed to the river of fish streaming far above our heads. "Isn't whatever can do that more likely to be what keeps the river rising than random damage followed by survival of the fittest?"

The speaker frowned. "Are you a creationist?" he asked me. Up went the cry, "Creationist, Creationist," and they tried to hustle me out of the meeting.

"Wait a moment," I said, and I turned to the bystanders. "It's not just fish up there. We're up there too, we're evolved just like those fish. Whatever can keep living creatures so far above the ground, won't finding that out tell us a whole lot more about ourselves?"

"Creationist, Creationist" the speakers continued shouting as they closed in around me. I looked up at the dark clouds above the "river." Shiny silvery fish were falling all around me. "Why," I said, "This evolutionary theory of yours, it's nothing but smoke and mirrors!" A fish landed on my head. I raised a hand to brush it away, and it rustled. It was a leaf.

Looking up I saw a carpet of crimson and gold leaves stretching away above me in all directions. I was sitting under a tree in my own garden at home, leaves drifting steadily down around me. "Darn," I said to myself, "it was only a dream. Now I'll never know how it ends."

From making up stories like this I've come to see natural selection and mutation as no more than a fairy story, like the moon being made of green cheese. I no longer need fancy arguments to prove they didn't work, it's just obvious. I want to say, "Can't you see how ludicrous they both are!" But that doesn't convince people, so I have to come up with arguments, I have to make a case, though to my mind that's treating mutation and natural selection with more dignity than they deserve.

Now let's now turn to some more "epicycles." Let's tackle the "competition" that supposedly drives natural selection and the epicycle invented to paper over the unnecessary problems it introduces, "kin selection."

Competition

Natural selection got its start in a famous argument between two 18th century gentlemen over what to do about the poor. "Help them? A waste of time!" said one, an economist. "No matter what resources you give them, they'll just keep breeding until they've used all up all those resources, and they'll end up no better off than they were before." Why should they be exceptions? he went on, and being an economist he came up with a mathematical formula "proving" that natural populations always grow faster than available resources ("as a population grows, the area it occupies will always grows faster than its circumference"). In his imagination this simple formula pitted individuals of each species perpetually in lethal competition with one another for scarce resources. Why should the human poor be any different?

Not a bad piece of reasoning. Except it isn't true. Individuals of most species are not constantly fighting one another for scarce resources. In fact, the only populations his formula applies to are human populations in huge cities like 19th Century London fed by food brought in from the surrounding countryside in cumbersome horse drawn wagons.

And that would have been that. Except, one person after another picked up this piece of misinformation and used it as an explanation for how evolution worked. That's where both Darwin and Alfred Wallace, Darwin's co-discoverer of natural selection, got the idea. Both of them concluded that incessant competition for scarce resources among "conspecifics"—creatures of the same species—would lead to "selection." Creatures that were even just slightly better adapted to their surroundings, just slightly more "fit," would have a slightly better chance of surviving that constant grueling competition to survive to reproduce and pass their characteristics on to future generations. Over long stretches of time, the result would be a slow but sure process of evolution.

Let's first dispose of the idea that natural populations are always pushing against the limits of the resources available to them. It just doesn't happen. When you go for a walk in the country, do you see badgers covering the ground from horizon to horizon, snarling at each other? If it was true would lions and tigers spend most of the day snoozing, as naturalists tell us they do? I don't see the deer in my backyard "competing" for my fallen apples. What sets limits to most natural populations is predators, disease and bad weather, most of which occur only now and then and in most cases either kill you or they don't. In between those kinds of bad luck most creatures find plenty of resources. "Conspecifics" do better by cooperating in dealing with predators and bad luck than by competing for every blade of grass under their feet.

Because both Darwin and Wallace were inspired by the idea that evolution resulted in constant competition between the members of a species, they once again created a problem where before there hadn't been one.

The problem was, how can you account for "altruism"— members of the same species being nice to one another— if they're in constant competition for scarce resources?

"Kin selection"

Time to introduce another epicycle. This time, it's "kin selection." According to natural selection, all that's needed for you to make a contribution to evolution is your fitter genes getting passed on to the next generation. It doesn't matter where those fitter genes come from. So if you do something smart that costs you your life but passes your genes on anyway, evolution's satisfied. The gene for whatever you did that was smart, even if it did cost you your life, will survive and appear in your descendents.

In practice, I can't see this working. Suppose you're a rabbit and you see a predator. By giving an alarm call that gets you eaten by that predator, you supposedly save enough of your relatives for the genes you share with them to get well represented through their progeny, instead of yours, in the next generation. This assumes, of course, that the predator is going to have time to chase down all your family members who all happen to be right next to you and who don't bother to run away. This could be true of predators that hunt in packs, like wolves, I suppose, but I think most predators usually kill and eat one prey animal at a time. Isn't that more likely to be the one that calls attention to itself while the rest flee to safety?

How about motherhood, surely the prime instance of altruism? In mammals, birds too, a mother frequently risks her life to save the lives of her young. Is that an example of kin selection at work? Not by my figuring. Consider a bird or mammal species where mothers have a lifetime average of ten young, and a female of that species having her first infant. Since on average only two of her ten infants will reach adulthood and reproduce, her first infant promises her genes only a 20% chance of being passed on, and she should favor her own survival much more than that of her infant. Here's why: mortality occurs most frequently in infancy and childhood, so she herself, having survived to adulthood, presents her genes with an 80% chance of being passed on through the remaining children she can expect to bear. According to the theory of kin selection you'd expect her genes to make her value her own survival over that of her young until she'd had about half her anticipated allowance, and they promised to pass as many of her genes on as she did herself. This is clearly not what happens.

Kin selection also fails to account for altruism between species, such as can occur between dogs and cats sharing a household, who surely have nothing to gain from their mutual affection. From a trustworthy source I heard of someone entering their kitchen at night to find their dog, with a mouse crouched between its paws, growling fiercely at a semicircle of watchful cats. Whatever the mechanism is behind the familiar experience of becoming friends with our pets, once we've identified it we'll no doubt be able to extend it to account for altruism among conspecifics.

The logic of kin selection does seem to apply among social insects. But that can be seen as simply the use of an appropriate tool by the Violent Engine. Elsewhere in nature kin selection isn't needed because natural selection isn't driven primarily by competition, despite formulas dreamed up by 18th century economists to win an argument.

Why selection isn't important for evolution

According to Darwin, who I trust on matters like this, female elephants on average have a total of 6 young. In other words, only four are lost for every pair that survives. But sometimes whole elephant families die from drought or disease. Except when entire families are lost like this, almost all elephants presumably survive. So having six young for every pair may be necessary to keep the numbers up without any selection for relative fitness going on. And while we're talking elephants, fossil records show that elephants that found their way onto islands millions of years ago shrank rapidly in size, as tends to happen to animals on islands. Their bodies decreased in size between 5% and 50% in fewer than 100 generations. Elephants aren't like images in Photoshop that you can shrink with a single command, there's no on-off size switch. This must involve a huge number of changes. Brought about by natural selection over fewer than 100 generations? I don't think so.

Take maple trees shedding seeds by the thousand. Why do they shed so many? So young maple seedlings will compete fiercely with each other, and the fittest get selected? More likely it's so there'll be enough for one or two to land on the occasional patch of suitable soil, wherever it is. Even then, if several seedlings grow up near one another, it's the one that gets the most light that will push up into the canopy. That's not competition, it's being in the right place in the right time. And maybe that's the main reason for most instances of generations consisting of large numbers of creatures: to cope with random variations in the environment so there'll be just enough progeny at the right place at the right time. Make that assumption and there may be very little selection through competition going on among members of the same species. Whatever drives evolution, it probably isn't competition among conspecifics as Darwin claimed.

If you use common sense for a moment, it's obvious you can't make new species by natural selection, any more than you can make new fashions by shopping. They're similar activities. In each case, you're selecting one thing, and rejecting something else. All you do by shopping is get store-keepers to bring out from storage more of the things you've already bought, to replace them. Until someone sits down with a pencil and paper to design something new, shopping doesn't create anything. And neither does natural selection. There's no designer of new stuff built into natural selection.

It's very simple. Natural selection is just selecting from what's available. That can't create anything new.

Who's clustering the genes?

Consider this—one moment you've got an ancestor of an elephant without a trunk, ten million years later elephants have trunks. The trunk's many functions work harmoniously together, a nice piece of engineering. How did the genes for the trunk all evolve at the same time and end up so coordinated?

We have two choices. Either all the genes coding for that trunk evolved through natural selection, getting sorted independently of one another and at random before being selected for, or they were somehow kept together and evolved as a unit. If they evolved through natural selection independently then it really is a miracle that they all ended up as so finely engineered a trunk. If they evolved together, then the fine engineering wouldn't be so surprising, but they couldn't have evolved by natural selection, they weren't resorted at random in each generation before being selected for. For another example think about whales evolving from a land animal, and how many adaptations to aquatic life had to take place simultaneously. Did all the genes for all those adaptations get sorted independently of each other in each generation, and prove worthy of selection independently of each other? I doubt it. Seems more likely they evolved as a unit.

Let's see what experts say about genes evolving together. It's talked about under the headings "clustering" and "gene order conservation." Here are extracts from a research paper:

> Even between very distant species, remnants of gene order conservation exist in the form of highly conserved clusters of genes. This suggests the existence of selective processes that maintain the organization of these regions.... The reasons for the maintenance of gene order are still not well understood, as the organization of the prokaryote genome into operons and lateral gene transfer cannot possibly account for all the instances of conservation found.

So no one's yet figured out how to account for the co-evolution of thousands of factors at the same time through natural selection, such as does occur and such as you'd need in land creatures while they evolved into whales.

The origin of species, still a mystery

We know there's a mechanism for new species appearing and old ones going extinct, because it happens. That's what I'm calling the Violent Engine. Can natural selection account for the origin of species?

Actually, despite Darwin calling his most famous book *On the Origin of Species...*, it doesn't account for how species appear. It wasn't meant to. All Darwin wanted to do was suggest a way creatures could become better adapted to their surroundings without God being involved. When they'd adapted enough, he assumed, they'd become a new species.

But what's most notable about species is, they don't just merge into one another continuously. Evolutionists like to point to exceptions, such as "ring" species where there is graduation like this, but it's the exception confirming the rule. The great mystery of species that so baffled the Victorians was, what keeps them distinct from each other, like separate creations? Why do they seem to stay the same for a million or so years and then, relatively suddenly, a new species appears and the old one goes extinct or one species splits into two.

That can't be accounted for by natural selection. Natural selection would cause gradual continuous change in a population. Another strike against it.

Wallace, co-discover of natural selection, turned against natural selection

Alfred Wallace was only 35 years old when he submitted the idea of natural selection for publication. Initially his theory and Darwin's were so alike Darwin said the headings in Wallace's account could just as well have come from his.

As an independent discoverer of natural selection, Wallace was very well qualified to question its capabilities. And question them he did. In just a few more years he came to the conclusion that natural selection couldn't be the only the mechanism behind evolution, because of one situation where it clearly couldn't do the job. That was the evolution of human talents.

Wallace wasn't a creationist. He had good scientific reasons that no one's been able to refute. Remember what I said about "disuse"? If creatures move to a new environment where some of their talents are no longer useful, natural selection no longer maintains them and they'll fade away from "disuse." The classic example is creatures with eyesight losing it when they start living in underground caves. Wallace saw this should apply to human talents as well.

Before I tell you what he concluded, I want you to realize that he was in a very good position to make the judgment he did. He was a man of two worlds. He grew up in 19th Century England. But he then lived for many years in South America and in Indonesia among people untouched by that kind of civilization. And he developed a great respect for them; he collected detailed information about their languages, for example. He really knew them.

Comparing their world with the world he grew up in, he became aware of a very obvious difference between the two. "Civilization" brought out talents such as mathematics, musical composition and artistic skills that people outside civilization made very little use of. In their world, as far as he could see, these talents made no contribution to survival. You might think he was mistaken but don't forget—he was one of the 19th century's foremost travelers, observers, and scientists. He observed a world that by the 20th century was fast disappearing. You probably couldn't make this kind of comparison today.

Suppose natural selection had built these talents into the human species, he said. If those talents weren't used, as his experience suggested they weren't, then natural selection would have got rid of them through "disuse," as I explained above. Those talents then wouldn't have been available for civilization to take advantage of when it came along. Since natural selection couldn't be what maintained those talents, there must be some other mechanism that did.

Then he went one step further. Maybe it was this other mechanism that generated those talents in the first place. If so, he speculated, we may have a lot more talents in us than we're aware of, that civilization hadn't yet tapped. This may be hard to imagine, but think of "idiot savants" whose only talent is figuring out instantly, for any conceivable date, what day of the week it falls on. That's a talent natural selection couldn't start selecting for until calendars got invented!

Wallace's theorizing opens up the exciting possibility that we could have still more untapped talents in us. But if you stay with natural selection, you'll never think of looking for them.

THE FINE PRINT
Benefits of a self-based science
See Part 2, Chapter 4.

This book is a manifesto challenging natural selection and calling for a new science. Once you set your mind to it, coming up with a new science isn't hard, as I demonstrated in Part 2, Chapter 4. First you set down your assumptions, then you work out their implications.

Do me a favor, though. Don't judge the rest of this book by the theory I came up with, by whether you think it works or not. You'd probably start off with a different set of assumptions and end up somewhere entirely different. That's fine. I hope you come up with something better. I was just showing how I go about it, starting with my own particular assumptions.

On the other hand, I will stick up for the kind of self-based science I came up with.

Background to a self-based science
To me, it just makes sense that thinking and evolving are two forms of the same thing. Once you say, consciousness isn't material, what else could it be except our thoughts evolving? What else is there? All I know exists in the world is matter, and selves. Our conscious experience, and evolution, if they're not just matter they must involve selves.

I know, it sounds hopelessly romantic! When I'm thinking, that's mental creatures evolving in my self! And when a new living species evolves, that's nature thinking!

Yet I do believe that. It's one of my starting assumptions. It makes sense of all my experience. Through conscious decision-making I can be genuinely creative, and that creativity can drive changes in brain chemistry and get expressed in the material world just as changes in genes get expressed in living creatures. And to me, living creatures are not the result of a physical process of adaptation, like wax being pressed against a mold. I see them as creative works of art. I mean that literally.

Imagine, instead of natural selection, a technology based on a self-based science. Imagine being able to account for everything you experience through a combination of evolution and material science. Imagine using the same word for "evolving" and for "thinking." Imagine using the same word for "thought" and for "species." Imagine being able to "evolve" your thoughts in new ways, or read the thinking in a landscape. Imagine understanding consciousness, by far the biggest challenge left in science with the most to tell us about the world that we don't yet know and the greatest potential for improving human life.

Imagine how grand such a self-based psychology would be. We'd have left far behind the cramped and squalid cabin of today's evolutionary psychology, with its narrow focus on sex and power, fear and greed. The self would borrow grandeur from the process of evolution, it would feel vast and rich, as vast and rich as nature itself. Our understanding of our own selves and our theories of evolution would keep leapfrogging each other to new heights of refinement.

On the other hand, I know that, judged by existing science, my proposed new science has huge problems. It supposes the genome has a self. Which genome? Where? How many genome selves are there? One? One for each species? For each person? For each cell? But the surfacing of these problems doesn't make the idea valueless. There's enough wrong with natural selection to justify exploring alternatives. And having two incompatible theories has served science well in the past, for example having both wave and particle theories of light, and both epigenetic and preformationist theories of development.

Here are a few speculations that can help paper over some of the gaps in my theory:

The world contains selves. We know this because we have the experience of being one. You can't (yet) construct this experience out of physics and chemistry.

Selves depend on matter. Judging from our own selves, selves must be associated with lumps of matter, in our case brains. These lumps of matter probably must be hubs for intense communications, and selves probably depend on persistent activity in the communication hubs that support them. Once that communication ceases, those selves vanish. That's true of our conscious selves, and it's probably true of individual genomes too.

Selves depend on a second kind of communication. I see no alternative to supposing there's some other means of communication, at present unknown, that acts as a carrier for the activity of the conscious self. If that's true, it may have to be true of a genome-self too.

Selves can be "creative." The activity of the conscious self is not limited by known laws of conservation or other principles of current physics. From our experience we know we can be truly creative. If evolution involves selves that could be true of evolution as well.

Genome as self service station. Because our selves last only a lifetime it seems likely to me that we get them from some persistent source. Since we're an evolved species and evolution shows signs of being creative, that's an obvious place to look. And as I wrote earlier, it's possible to regard the genome as the material support for a communications hub that's persisted for several billion years. We could think of the genome-self as a filling station we each pull up to for a charge at birth.

Genome as communication system. Development, homeostasis and evolution I'm assuming are all driven by selves of some kind, presumably by genome-selves. For chromosomes in individual cells to become units in a communication hub on the scale of the brain or larger, they must communicate with each other in some new way. Maybe some of the repeating "junk" in the genome is comparable to the girders making up a radio station's transmission tower. Maybe the genome is very largely a transmitting and receiving apparatus.

Genes as memory. Thinking along these lines, of the genome as serving as a brain, suddenly three billion base pairs seems barely enough. There is nowhere else but in the gene pool for the creator of our selves to keep its memory, or its information about the current state of species, or its account of the physical world to which it must fit its creatures. The genome must be primarily memory, then some transmission and receiving apparatus, with very little left over for transcribing proteins and prescribing growth patterns.

Let's take stock

Much of what I'm proposing is not really that far-fetched. It's not that far-fetched to suppose that the genome consists of memory and communication apparatus as well as a blueprint for development. It's not that far-fetched to imagine a genome self being the source of our selves. It's not that far-fetched to suppose the interaction between matter and consciousness goes both ways, as touch, sight and hearing in one direction and back in the other as speech, writing and body language. In fact, none of my suggestions is all that far-fetched except for the new communication system I'm forced to propose as a carrier for selves and their activities, in which would lie the secret to conscious experience, conscious decision-making, dreaming and free-will, and the management of development and homeostasis and the evolution of new species, new organs and tissues.

Imagine that communication system being discovered. Now the rest of what I've proposed would likely follow. The Two Cultures would embrace one another, science and history and art would join in a new "consilience." Science's past four centuries would seem strange indeed.

Missing from such a new science would be mutation, genetic drift, and all natural selection's other epicycles. Also missing might be adaptation and variation. In their place might be anthropomorphisms such as: find out, try, decide, and create.

Experiments in self-based science

While that new system of communication remains to be discovered, there are other ways to anticipate a self-based science. Here are some suggestions:

1. *The resorting of genes at meiosis*: is it random or does it involve intelligence? Experiment: set the process a problem such as the mating of a horse and mule. If the process is random, it should always take the same time. If intelligence is involved, if there's a problem to be solved, it may take longer.

2. *Can pre-adaptation be proved to have taken place?* Natural selection can't account for that, a self-based science can.

3. *Non-adaptive variation.* If some inherited behaviors could not conceivably have ever been adaptive, such as pigeons flipping somersaults, that might prove that variation is not the result of natural selection.

4. *Identifying distinctions between natural selection and the genome self.* They could be hard to tell apart. Both can account for creatures becoming better adapted to the environment. Both can account for where variation comes from—"mutation" according to one theory, according to the other the genome-self's thoughts driving changes in genes. Both could involve trial and error, though here there's a big difference: the genome-self can use intelligent trial and error, natural selection is really nothing more than trial and error at random. But when analyzed statistically the results are likely to look similar and all be branded as "mutation."

   The biggest difference, I think, lies in form. Living creatures may display more form than adaptation requires. If excess of form could be measured and is significant, that would favor a self-based science.

5. *Evolutionary processes.* If evolutionary processes and our thinking run on similar lines, then examples of evolution could be looked at for processes such as induction, deduction and logic that we use. Some people claim natural selection can mirror the results of such processes, given enough time, but that's like saying random typing will eventually reproduce all the works of Shakespeare. In practice, that kind of random trial and error should be distinguishable from intelligent strategies.

Afterword
AUTOBIOGRAPHY OF AN EVOLVED SELF

My earliest memories center around an old and tattered children's encyclopedia. Its black-and-white photos recalled the heyday of the Victorian period. From it I learned about the marvels of "modern" industry and science. How pins are made. Great inventions such as railways and steamships. Major scientific discoveries. Poring over it, I dreamt of discovering how everything worked, even myself.

I came to think of myself as Mechanical Boy.

Mechanical Boy found it natural to experiment on himself as if he was a machine. I took up 3D photography. This involved taking one photo of something, sliding the camera a bit to one side and taking a second photo, then putting the two photos side by side and viewing them so the images superimposed. It was presenting my eyes with photographic versions of the views they were used to using to create 3 dimensional space.

I played with my eyes as if they were optical instruments on a workbench. By taking two photos with the camera moved only a quarter of an inch in between, I could make a toy train look like a giant locomotive. By taking photos on board ship a few seconds apart I could see the shoreline and clouds in exaggerated depth. I took photos of slides pivoted on a knitting needle to give me views from two different directions into the microscopic world. The more I treated my eyes as instruments, the more mechanical I seemed.

I've since learned that, just a few miles away, just about

then, Oliver Sacks was doing much the same. Maybe this is an English thing: I discovered that I shared another experience, exploring the will, with the British philosopher Brian Magee. Like him, I would hold my finger straight and say, silently to myself, "Bend." But it wouldn't. Just saying it wasn't enough. Not until I willed the finger to bend did it actually bend. The saying it and the willing it both happened inside the self. But they were different. I puzzled over that difference. What technology lay behind them—behind the saying something to oneself, and willing it?

Later, as a teenager, Mechanical Boy stumbled across a book very different from his ancient wonders-of-technology encyclopedia, a book that offered him a very different kind of explanation.

My father was a Church of England clergyman for an outer suburb of London. Every Sunday I went to church and sang in the choir. On Saturdays I acted as usher for weddings. My bedroom window looked directly out onto the church.

Curiously, the household was not religious. We didn't say prayers together, and I wasn't pressed to relinquish my dog-eared encyclopedia for the Word of God. I was never a believing Christian, nor asked to be one. I never converted to religion, so I never needed to rebel against it.

While my father wrote his sermons downstairs, I would be upstairs in my bedroom practicing my hobbies. For example, from a dim recollection of an illustration of early clockwork I made a ticking timer for darkroom photography with an escapement mechanism made of springy plastic collar stiffeners engaging a gear wheel made of 60 pins pressed into the edge of a balsa-wood disk. One spring, when I was in my mid teens, I embarked on the hobby of becoming an orator. I would stand in front of my full-length mirror and, while the church steeple loomed outside my bedroom window, declaim to myself, aloud.

The book I chose, for its grandiose style, was Darwin's *The Origin of Species*. I ended up extracting from it more substance than style. In the course of a few weeks reading aloud to myself in the mirror, I came to know myself as an evolved creature, through and through. Even today, if you could prove to me that evolution isn't true—that I wasn't evolved—I would still have to believe in it because, from that early experience, evolution became the basis of all my later beliefs.

As I closed Darwin's great book, I realized the explanations I wanted for my self wouldn't come from physics and chemistry alone. They'd come from evolution. Not so much evolution of the body, that didn't seem much of a mystery. The exciting part would be learning where consciousness and thoughts came from.

To learn more about his self, Mechanical Boy went to University College London to study biochemistry. I learned how my body extracted energy from food and used it to make my muscles contract and power my brain. I learned a lot more about myself as engineering. But I didn't learn anything about the evolution of the self. I never could get with the program. I failed my degree and, answering an ad for someone "insufficiently specialist-minded to obtain their degree" (i.e., educated but cheap), entered the world of book publishing as a designer.

At the age of 29 Mechanical Boy left London for New York City, and a severe case of culture shock. He found himself having to cope with American-style individuality and self-expression. I remember taking an evening course in creativity. The instructor asked us to go outside the classroom and line up in the order of how creative we were, less creative to the right, more creative to the left. Thinking like an English person, I steeled myself for the inevitable jostle to be in the middle, right opposite the door. To my surprise I found myself alone; the entire class was hurtling away from me in the direction of greater creativity.

People were much more interested in being creative than seeing themselves as machinery. For example, I grumbled about how Americans talked. Conversation in America didn't seem to be about anything particular. People didn't obey the conventions I was used to. So, to educate Americans in what I thought was good conversation, I invented a conversation-game, complete with a points system and rules. To my surprise, I found few takers. Americans didn't want to treat conversation like a piece of machinery they could take apart and put together again. They wanted to tell each other stories.

Somewhere around here appeared Teilhard de Chardin's book *The Phenomenon of Man*. He proposed "The Omega Point," a point far in the future we were drawn to, a destiny, when human evolution would reach its goal of something-or-other. For me, this defined exactly what I wasn't interested in. I wasn't interested in any future goal. What I wanted to know was how a process without goals, like evolution, could get us where we are today. In fact, playing my conversation game its one and only time, my partner and I discovered that we felt differently about evolution for just this reason: he saw evolution as a drawing on from in front, I saw it as a pushing from behind.

By this time, in his mid thirties, Mechanical Boy had developed a very strange way of looking at the world. This strange point of view had matured at just about the time celebrated in his childhood encyclopedia, back in the late

19th century. It fitted right in with Mechanical Boy's view of his self. The technical term for it is "epiphenomenalism."

If you're a regular person, not a scientist, you'll probably be astonished anyone could for a second believe in something so obviously lame. It says, you can't do anything "consciously." Everything you do is controlled entirely by your brain; consciousness is just something the brain gives off, like your shadow being cast on the ground by your body. Like your shadow, consciousness has no power to direct you, to influence what you do. Everything you do is driven entirely by chemical processes taking place in your brain.

Mechanical Boy actually believed this. He enjoyed monitoring his conscious thoughts, while taking it for granted that they had no influence on his behavior. His actual decisions and actions were due entirely to physics and chemistry in his brain. Other people's assumption that their decisions came from their conscious thoughts was just, he assumed, a delusion.

Then, one day, he realized he was wrong. I could talk about consciousness. I could write about it. In fact, there wasn't any aspect of this supposedly inaccessible subject I couldn't talk and write about. And speaking and writing are clearly things happenings in the physical world. Through other people hearing and reading about my conscious experiences, I could affect their consciousnesses, and what they said and did.

Then I realized I was expressing my consciousness in matter all the time, through facial expressions, gestures, drawings, decisions I made. All the time my consciousness was giving off clouds of physical effects. You couldn't help reading my consciousness just by being near me and automatically noticing these effects. And this was true of everyone. I was picking up other people's consciousness the same way.

This hit me like a lightning bolt. Something that most people take for granted and don't give a second thought to, became for me absolutely extraordinary—mind can interact with matter. Not only can the physical world act on our consciousness—we can experience seeing and hearing it, for example—but our consciousness can also act back on the brain to affect the physical world. It's happening all the time, all around us. It's in our architecture, it's in the litter lying by the side of the road, it's in every gesture and every sound we make.

This was absolutely certain. Yet science seemed to have nothing to say about it, as if there was nothing particularly interesting about it. Well, I found quite a lot to say about it, about how mind and matter interact. If every now and then you find my conclusions a little far-fetched, remember, I'm coming from a very special place—I used to be an epiphenomenalist.

In 1972, after five years as a book designer for Dover Publications in Manhattan, I left to write a book. I didn't at first know what book. It began as an early crack at the evolved self theme, then became "Our Little Differences" about how much people differ from one another both physically and mentally—Mechanical Boy was puzzled that machines differing from one another as much as people did could still function OK. Finally a new discovery took me in yet another direction. I realized that conversation had to be a human invention. It arose too recently to be evolved. You couldn't have a conversation until language had been developed, which seemed to have happened only about 50,000 years ago. And from my own experience of British and American conversation I knew that different people did it differently. Here was something that seemed to be part of nature but was actually part of culture. I started seeing consciousness as a creation of both evolution and history, they were not so separate after all.

After five years of not finishing or publishing anything, with no end in sight, I returned to "work," this time as a scientific and medical writer. But I had finally learned something about how the self is put together.

In my early fifties, I began having experiences of ecstasy. They'd come on slowly, with a feeling of something enormously significant about to happen, then deepen until I became transfixed by the wonder of whatever my eyes fell on. Once, while I sat in my car, it was a group of men working on a gantry against the wall of a distant building. The experience at its most intense lasted about ten minutes, then gradually faded. It didn't feel like a religious experience, though afterwards I would have a feeling of great wonder and gratitude. I remember saying to myself, even one of these experiences could make you feel your entire life was worthwhile.

The last of these experiences was the most remarkable. It came on while I was strolling through a country fair. As the experience deepened I began consciously turning my attention on first this aspect of the scene, then another. Whatever I turned my attention on became illuminated as if by a searchlight, and I could "read" it as if my vision and understanding had become superhuman.

I realized I was sharing the experience of innumerable saints and heretics, who attributed it to all kinds of gods and demons. So I asked, distinctly, within myself, Who is this? And the answer came back, as clear as if spoken, "This is me." And it was clearly a part of me that spoke, back to myself. I was thrilled. This wonderful way of seeing was part of my nature. The path to this exquisite wisdom, this rapturous experience, lay within me.

This experience revealed that I carried buried within me advanced mental powers that usually I was unaware of. Where did these powers come from? Since I'm evolved, presumably they had evolved too. Maybe by studying how evolution worked I could tap these powers in myself. I began to take evolution more seriously as the source of a power present within us.

Sometimes, the book you need just shows up. For me, at this time, it was Julian Jaynes *The Origin of Consciousness in the Breakdown of the Bicameral Mind*. Very popular with us laypeople, a Book of the Month club non-fiction bestseller, it remains almost uncited in the scientific literature. I regard as a major personal loss that Jaynes never published the second volume he promised, covering the past 2000 years.

Having early on graduated from physics to evolution in a search of the answers to life's big questions, Mechanical Boy was now looking beyond evolution to history. From having been exempted from taking the "O-level" history exam so he wouldn't disgrace himself, history now merged with biological evolution to become his primary hunting grounds for leads to the engineering behind the self.

What kind of a thing was this "self," that physics and chemistry, evolution and history together had engineered? How "big" was it? How "complex"? I did have a model. As I said before, although my father was a clergyman there wasn't much Christianity in our home. Perhaps because of that I never rebelled against the church. Instead I became curious about it. People must have a good reason for coming to church, I thought. But one by one, reasons for believing in Christianity dropped away. I was impressed that people had experiences of religious conversion. Then I read "Battle for the Mind" by William Sargant and learned that religious conversion was very similar to political brainwashing. I remained impressed by mystics' reports of ecstasy—until I had them myself.

But I look to religion for a special kind of wisdom about human nature. Once you create supernatural beings you can make them do anything you want. You can use them to explain anything, or make anything happen. So over time religion becomes a record of what people most want, what they most need explained. Religions are plumbings of the most profound depths of human nature. A good story about human nature should provide all the satisfactions and explanations of all the world's religions. Why settle for anything less?

Yet we do settle for less. When I first caught a glimpse of how powerful evolution could be, I began reading books on animal behavior. I hoped they'd give me clues to where my consciousness and thinking came from. But they didn't. I never lost hope, though—surely some of those authors would move on to human behavior and come up with the answers. But I waited in vain. Eventually sociobiology came along, then evolutionary psychology, but even they didn't answer the deeper questions. Who knows, it might be centuries before science comes up with the answers we want. If we were going to find any kinds of answers in our own time, they'd have to come from extensions of what we already know.

In some ways, I remain Mechanical Boy. I still like to experience my senses as instruments and experiment with them. When I step outside my front door in the morning to retrieve the newspaper I savor the impact of the outdoors on my various senses, perhaps dwelling on the sight of snow on the ground and the sound of my feet crushing it, in combination with the feel of the impact of the ground on my feet. Freedom and will have concentrated down to this one essence—being able to consciously direct my attention. For me that is the ultimate exercise of freedom.

By harnessing the senses this way, I've turned washing up into a meditation on the self. First I make sure there's some seed in the bird feeder outside the window. Then I put on some music very loud, and just listen, close up between the two speakers. Finally I turn from the speakers and begin washing up, slowly and deliberately, as a form of worship of things. Sometimes when I see something particularly wonderful in the sink I photograph it. I have a collection of about 30 such photos, some of which I've exhibited. All very Zen. But Western and scientific in inspiration.

I titled a show of these photographs, "All Washed up in The Hudson Valley," which is where I live. Punning intrigues me. "I hear Ireland's booming," said my wife. "Oh," I said. "You can hear it from here?" "What do you think of the cloning of farm animals?" she asked me. "I'm in two minds about it," I said, "but they're both <u>exactly</u> the same." We've had language of any kind for only a few tens of thousands of years? Where could such an advanced capability as punning come from?

Americans and Britons seem to have different kinds of self. That's probably due primarily to the distinctively British school of philosophy known as the Cambridge Footlights. The relentless advance of modernism started being beaten back in the 20's by Wittgenstein at Cambridge University. He returned there to teach in the 30's, where I am convinced his personal style became imprinted on the

University, and eventually on all of England.

Wittgenstein is pivotal in the development of the late-modern British self, I'm going to claim. I view him as an involuntary solipsist, perpetually terrorized by the instability of the world as reported by his senses, every second threatening to dissolve into chaos. He could hold it stable and comprehensible only momentarily through extreme acts of will, but whenever he relented the chaos would swoop back and overwhelm him. In his first book he attempted to make logic force the world to remain stable. But the attempt was a failure, as he says at the end. On his return to Cambridge he developed instead the idea that we make the world stable only through games, only by making up sets of rules that we all agree on. That was the only way he could imagine us sharing the same meanings of the world. Ontology and epistemology were just agreements we made up, that defined our stock of meanings.

Britain was softened up for Wittgenstein by Spike Milligan, writer for *The Goon Show*, himself only marginally sane. I was just pre-teens when *The Goon Show* rocked England. Over a decade or more it laid the foundation for later British humor. Then, out of Cambridge poured one generation after another of graduates of the Cambridge Footlights theatrical society, first Richard Cooke and Dudley Moore, then "I'm sorry I'll read that again," leading up to the ultimate expositors of Wittgensteinian games, Monty Python. Here, from memory, is a sample.

"I'd like an argument please."

"Certainly Sir, that will be 5 pounds. Up the stairs and on your right."

"Thanks."

Walks up the stairs and in the first door on the right. Is greeted by:

"You blithering idiot, what do you think you're doing standing there...!"

'Here, that's not an argument!"

"Oh, sorry, this is cheap personal abuse. Argument's next door.

"OH, sorry, right."

Then the skit takes apart what is or is not an argument.

"Time's up."

"No it isn't."

"Yes, it is,"

"No, it isn't.

"Yes it is."

"Look, this isn't an argument."

"Yes it is."

"No, it isn't."

Saturday Night Live seems very tame when you've been trained in epistemological uncertainty by Wittgenstein and his followers. Abuse of people in power is trivial compared to abuse of the fundamental principles underlying our shared sense of reality.

Once trained to question those principles on a weekly basis, one is prone to speculate, particularly about the self, which no longer has any hard and fast principles holding it together.

The aspect of our shared reality I have my sights focused on most tightly is natural selection. How do advocates of natural selection account for talents of the self such as humor? They are likely to say, humor is an adaptation. Pressed for more detail, they will probably say it's a product of sexual selection—it demonstrates quickness of mind, which would make jesters more attractive to potential mates and so increase their chances of passing on their genes. In matters of the self, natural selection has become that odium of science, a hypothesis that accounts for everything while explaining nothing.

Darwin started out as a hero for me, as a scientist, as a thinker, and as a person. But over time I have lost some of that hero-worship. An observer and experimenter of genius, yes, but less of a thinker. It appeared in his time that, for natural selection to work, inheritance had to come as particles that didn't blend. Darwin came up with a theory he called "Pangenesis." But it wasn't very good and when his cousin proved it couldn't work the way Darwin proposed, Darwin just sulked, he wouldn't do the hard work of figuring out how these "gemmules"—as he called these early intimations of genes—really worked. Seeing how flawed his thinking was in this instance brings into question his idea of natural selection. Wallace, who was the first to submit it for publication, a few years later began to doubt it, for good reason, and to qualify it. But Darwin's response was only a wishy-washy entreaty to not "murder our brain-child," something like that. Natural selection was an idea obvious enough to be "discovered" at least five times in the 19th century, it didn't take a genius. And the more I considered it the more I found it inadequate, at first as an account of the origin of the self as I discovered that in me, then as an account of anything else.

I have labored at coming up with an alternative. My laboratory, my mental discipline, has been writing. The process of writing operates on one's initially confused thinking like a spinner eliciting a continuous thread from a bundle of lamb's wool. I began in 1992 recapitulating in the form of a Utopian novel the implications for human nature of the stages in evolutionary theory's advance over the past couple of centuries (*Father, in a Far Distant Past, I Find You*). Starting in 1996 I began work on a self-help manual based on evolutionary principles (unpublished). Clues gathered through writing that book led to the writing of a light romantic novel introducing the issues I perceived lying behind the natural selection/creationism controversy (*Me and The Genies*). Ideas generated by writing that book provided the basis for the anti-natural selection manifesto *Save Our Selves from science gone wrong*. In what I like to think is a Wittgensteinian way, I first point out what I think is wrong with our current evolution game, then offer a new one.

A self based on scientific wisdom should be as grand as the religious self ever was, and more. That vision of a self-based science, supporting at least the self I am now, and the greater self I can only dream of being, will be the theme of future books.

Recommended readings, notes and sources

OTHER BOOKS COVERING THIS BOOK'S MAIN THEMES

History of ideas associated with science and evolution

An excellent introduction to scientific thinking from the Ancient World on. Anthony Gottlieb, (2000). *The Dream of Reason: A History of Western Philosophy from the Greeks to the Renaissance*. W. W. Norton & Company.

For a variant reading of this history emphasizing the religious origins of science, David Noble, (1999). *The Religion of Technology: The Divinity of Man and the Spirit of Invention*. Penguin Books.

A wonderful, extremely readable account of people and events associated with the founding of the Royal Society. Carl Zimmer, (2004). *Soul Made Flesh: The Discovery of the Brain and How it Changed the World*. Free Press.

Bacon's classic is an extraordinary window into how confused, by modern standards, even the best informed intellects were about science four centuries ago. This is the program inspiring the Royal Society and laying out the problems Newton became most famous for pursuing. Together with Zimmer it provides an amazingly rich sense of the context for the founding of The Royal Society. Francis Bacon, (1620). *The New Organon*. Ed. Lisa Jardine, Michael Silverthorne, (2000). Cambridge University Press.

For Positivism I recommend checking out both Positivism and August Comte in Wikipedia. The classic reference is John Stuart Mill (1865). *August Comte and Positivism*.

I first learned about history of science 50 years ago from a Penguin book by Stephen Toulmin. By 1989 he had come to reject modernism and preached a return to Renaissance humanism. Stephen Toulmin, (1989). *Cosmopolis: the Hidden Agenda of Modernity*. The University of Chicago Press.

Conciousness

THE essential reference. I no longer go along with everything he concludes, but he defines the ground. Julian Jaynes, (1990). *The Origin of Consciousness in the Breakdown of the Bicameral Mind*. Houghton Mifflin Company.

The main hot-head denying the existence of conscious decision-making. "Although it is not absolutely mandatory, most advocates of the scientific image assume that all thought and action is determined." And, "Freewill is disbelieved within mind science." Flanagan, Professor of Philosophy at Duke University, assures his students (undergrad psychology students among them) and us that we are determined, we do not have free will. "....it remains bewildering to me that, when pressed, budding mind scientists will acknowledge that they assume determinism in their practice as mind scientists, but they also don't assume it could really be true....Some mind scientists clearly see that this position is instable. But I am not sure they see clearly why if the traditional conception of free will is not a credible assumption in the lab, it is not a credible assumption outside the lab either." Owen Flanagan, (2002). *The Problem of the Soul: Two Visions of Mind and How to Reconcile Them*. Basic Books.

The main sober-sides denying the existence of conscious decision-making. To me, his books read more like consciousness explained away. Daniel Dennett, (1991). *Consciousness Explained*. Little, Brown and Co.

A dense and heavy textbook mainly detailing the failure of other scholars like those two above to deal with conscious experience adequately. Not an account of consciousness. David J. Chalmers, (1996). *The Conscious Mind: In Search of a Fundamental Theory*. Oxford University Press.

Stephen Budiansky, (1998). *If a Lion Could Talk: Animal Intelligence and the Evolution of Consciousness*. The Free Press.

Evolution

An account of evolution traced backwards from us, with times in millions of years. An excellent overview of evolution with an emphasis on our ancestors. Richard Dawkins (2004). *The Ancestor's Tale: A Pilgrimage to the Dawn of Evolution*. Boston: Houghton Mifflin Company.

Creatures' expressions of themselves in the material world interpreted as forming part of their phenotype. Human culture? Also introduces "memes." Richard Dawkins, (1982). *The Extended Phenotype: The Long Reach of the Gene*. Oxford University Press.

Recent discoveries on how form in living creatures is organized and coded for in the genome, further insight into the machinery of evolution. Sean B. Carroll, (2005). *Endless Forms Most Beautiful: The New Science of Evo Devo and the Making of the Animal Kingdom*. W. W. Norton & Company.

An extended and fully referenced critique of natural selection, proposing in its place an emergence theory of evolution and what the author calls "natural experiments." Robert G. B. Reid, (2007). *Biological Emergences: Evolution by Natural Experiment*. Part of the Vienna Series in Theoretical Biology, The MIT Press.

Highly-informed criticism of natural selection by a renowned science journalist, excellent body-blows delivered in the final chapter, good parade of counter examples in the middle chapters, worthwhile reading throughout. Gordon Rattray Taylor, (1983). *The Great Evolution Mystery*. Harper & Row Publishers.

Another critic of natural selection, good on Malthus. David Stove, (1995). *Darwinian fairytales: Selfish Genes, Errors of Heredity, and Other Fables of Evolution*. New York: Encounter Books.

A model for how to think and talk about evolution, and the kinds of things to talk about. Free of jargon, open to personal experience, courteous, engaging. Stephen Jay Gould, (1978). *Ever Since Darwin: Reflections in Natural History*. Burnett Books/Andre Deutsch Limited.

Darwin

Excellent life of Darwin and his circle, fairly critical. Adrian Desmond, James Moore (1991). *Darwin*. London, Michael Joseph.

Essays assessing how Darwin's ideas interacted with his and later times. *The Cambridge Companion to Darwin*. Ed. Jonathan Hodge, Gregory Radick, (2003). Cambridge University Press.

This kind of anthology is an excellent way to sample Darwin's personal character and his genius as a naturalist and observer. Ed. Duncan M. Porter, Peter W. Graham, (1993). *The Portable Darwin*, Penguin Books.

The master, gingerly revealing scraps about himself. Not his best work. *The Autobiography of Charles Darwin*.

Alfred Wallace

Wallace, brilliant naturalist, socially awkward, a natural socialist, had the courage to carry his search for the source of human talents into spiritualism, which in Victorian times had some of the promise the Human Genome project has today. Perhaps he deserves to be forgiven for this.

Andrew Berry (2002). *Infinite Tropics: An Alfred Wallace Anthology*. London: Verso.

Peter Raby (2001). *Alfred Russel Wallace: A Life*. Princeton University Press.

NOTES AND SOURCES FOR THE TEXT

by part and chapter, in order of appearance.

PART 1. How present-day science threatens the self

Chapter 1. The self in danger

"According to one authority of consciousness ..."
David J. Chalmers, (1996). *The Conscious Mind: In Search of a Fundamental Theory.* Oxford University Press, p. xiii.
"Our minds are just what our brains..." Daniel C.
Dennett, *Freedom Evolves,* Viking, New York, pp xi, 2.
Here is how a friend of mine stated the imperative to teach physicalism in the classroom: "Scientists do have a specific metaphysical tradition, right or wrong, and it IS material monism, and that is what we all assume and require is taught when the theories are taught."

Chapter 2. How science came to deny the self

"Ancient Roman magic was being revived." I'm referring to the hermetic tradition as described in Francis A. Yates, (1964). *Giordano Bruno and the Hermetic Tradition.* The University of Chicago Press.
For this chapter I draw heavily on David Noble, (1999). *The Religion of Technology: The Divinity of Man and the Spirit of Invention.* Penguin Books.
For the composition and activities of the Royal Society see Robert K. Merton, (1970). *Science Technology and Society in Seventeenth Century England.* New York: Howard Fertig.
"The King of England was actually persuaded..." W.
Bro Alex Davidson, "Freemasonry, The Royal Society, and the Age of Discovery." I downloaded this article September 13, 2006 from www.rsnz.org/news/venus/freemason_bg.php.
Francis Bacon, (1620). *The New Organon.* Ed. Lisa Jardine, Michael Silverthorne, (2000). Cambridge University Press.
Very good for this period is Carl Zimmer, (204). *Soul Made Flesh: The Discovery of the Brain and How it Changed the World.* Free Press.
"I found clues in a recent account of Freemason..." W.

L. Wilmhurst,(1927). *The Meaning of Masonry*. New York, Bell Publishing Company.

"Masonry and the Royal Society didn't separate..." See above "Freemasonry, The Royal Society, and the Age of Discovery."

"Darwin's grandfather Erasmus Darwin..." From *Wikipedia*.

"The object of worship of this new religion..." John Stuart Mill, (1865). *August Comte and Positivism*. I downloaded the text from Project Gutenberg, www.gutenberg.org.

For earlier Positivism, I mined this dense philosophical text. Robert C. Scharff, (1995). *Comte After Positivism*. Cambridge University Press.

"the positive [stage] is destined finally to..." John Stuart Mill's interpretation of Comte's intent.

"...ghost in the machine" Gilbert Ryle, (1949). *The Concept of Mind*.

"There's a movement in philosophy towards 'post-Positivism,'..." From Scharff, above.

Chapter 3. At the core of the threat: natural selection

"According to an article in the New York Times ..." David Brooks, *(April 15, 12007)* "The Age of Darwin," *New York Times*. New York.

"And not just in the US." "In the beginning" *The Economist* (April 21, 2007), p. 23.

Chapter 4. The "Violent Engine" hypothesis

I borrow legitimacy for reading this kind of meaning into my anagram from Carl Sagan's novel *Contact* (1997). Pocket.

I've read that Aristotle's term for a movement of an element contrary to its nature is translated into English as "violent." So a movement of fire downwards, contrary to its nature, would be "violent," as would any non-circular movement of a heavenly body. So "violent engine" nicely conjures up an engine acting contrary to the usual nature of machines. Perfect!

Chapter 5. Natural selection fails the "Violent Engine" test
"the young Charles Darwin came across a preliminary outline of Positivism..." Most information about Darwin in this chapter comes from *Adrian Desmond, James Moore (1991). Darwin.* London, Michael Joseph.

"From reading about Darwin, it seems he wasn't..." The editors may be surprised, but I came to this conclusion mainly by reading Ed. Jonathan Hodge, Gregory Radick, (2003). *The Cambridge Companion to Darwin.* Cambridge University Press.

"he came up with a theory he called "pangenesis..." Charles Darwin, (1868). "Provisional Hypothesis of Pangenesis," *The Variation of Animals and Plants under Domestication.* My information came mainly from comments about this in the above two books.

"Two other people had come up with the idea..." Ed. Duncan M. Porter, Peter W. Graham, (1993). *The Portable Darwin,* Penguin Books, p. 216.

"Darwin became the figurehead of a..." Desmond and Moore, above.

"In 1930, a guy called Fisher ..." I've not read this, but apparently no one can. I made do with scraps of information about it from various online sources. Ronald Fisher, (1930). *The Genetical Theory of Natural Selection.*

Chapter 6. Collapse of the epicycles
"Natural selection got its start in..." For a good critique of Malthus see David Stove, (1995). *Darwinian fairytales: Selfish Genes, Errors of Heredity, and Other Fables of Evolution.* New York: Encounter Books.

Chapter 7: When natural selection fails, Physicalism fails
"Why does a theory like Associationism fall..." Go

Steven! One down, one to go! Steven Pinker, (2003).
The Blank Slate: The Modern Denial of Human Nature.
Penguin.

PART 2. Blueprint for a new self-based science
Chapter 1. Questionable assumptions well worth abandoning
"In a book on facial expressions..." Charles Darwin,
(1872). *The Expression of the Emotions in Man and
Animals.*
"Can you convert one into the other mathematically?"
These two skulls appear to differ by far more than the local
changes in scale illustrated in D'Arcy Wentworth Thompson,
(1917). *On Growth and Form.*

Chapter 2. Attributes of "The Violent Engine"
**"Let's start by doing what Darwin's mentor
Sedgwick..."** Sedgwick complained that Darwin had
the process of scientific discovery upside down. It should
have the form of a pyramid, with an examination of all the
phenomena of nature at the base and a narrowing spire
of deductions emerging from it, until you arrive at the
mechanism at its peak. Darwin had instead arrived at his
mechanism first, then piled on top of it a growing mountain
of phenomena of nature cherry-picked to confirm it. From
The Cambridge Companion to Darwin.

Chapter 3. Candidate for the role of "Violent Engine"
"Examples are the exons and introns..." Eva Jablonka
and Marion J. Lamb (2005). *Evolution in Four Dimensions:
Genetic, Epigenetic, Behavioral and Symbolic Variation in
the History of Life.* The MIT Press.

Chapter 4. Source of selfs
"One of my cats is black..." Reading meaning into creatures
of another species is very problematic. It's easy to mistake

learned responses for evidence of a self. A good guide here is Stephen Budiansky, (1998). *If a Lion Could Talk: Animal Intelligence and the Evolution of Consciousness.* The Free Press.

PART 3. Manual: How to save the self

Chapter 1. Threat of a new barbarism

"It happened 4000 years ago..." Julian Jaynes, (1990). *The Origin of Consciousness in the Breakdown of the Bicameral Mind.* Houghton Mifflin Company.

"There's a new literary criticism..." "Next to sex and property, fidelity to kin presents itself as an urgent motivational force." In isolating this quote I am no doubt being grossly unfair to this source; I admit I couldn't follow it. Joseph Carroll "Human Nature and Literary Meaning: A Theoretical Model Illustrated with a Critique of Pride and Prejudice" in Ed. Jonathan Gottschall and David Sloan Wilson (2005). *The Literary Animal: Evolution and the Nature of Narrative.* Evanston, Illinois, Northwestern University Press. See also, for a reduction of the life of Darwin to the five principles I listed, Robert Wright, (1994). *The Moral Animal: Evolutionary Psychology and Everyday Life.* New York, Pantheon Books.

"Another strand comes from game theory." Matt Ridley, (1998). *The Origins of Virtue: Human Instincts and the Evolution of Cooperation.* Penguin Books.

"Even those feelings have to explained away." Joshua M. Ackerman et al, "Is friendship akin to kinship", *Evolution and Human Behavior,"* 28/5, September 2007.

"This story is spilling out into self-help books..." Terry Burnham and Jay Phelan, (2001) *Mean Genes: From Sex to Money to Food. Taming our Primal Instincts.* Penguin Books.

Chapter 2. Tactics for attacking natural selection

"How about movies?" Book: Christopher Buckley, (1994).

Thank you For Smoking. Random House. Movie: Released
April 2006, available on DVD.
"As soon as you speak out against natural selection"
Shaun Johnston, (2006). *Me and The Genies.* Evolved Self
Publishing.

Chapter 3. Promoting a self-based science
Primary theme: identifying adaptation with knowledge.
Thomas Huxley quoted (1881), "The struggle for existence
holds as much in the intellectual world as in the physical
world. A theory is a species of thinking, and its right to
exist is coextensive with its power of resisting extinction by
its rivals." Evolutionary epistemologists believe "learning
and intelligence are evolutionary processes embodied
in brain mechanisms. This means that whatever those
principles are that describe the evolutionary processes
operating between animals in the conventional sense of
evolution must be the same as those that operate inside
our heads." And, presumably, vice versa. Henry Plotkin,
(1994). *Darwin Machines and the Nature of Knowledge.*
Harvard University Press.
For a sampling of scientists converging on consciousness from
various directions, see *Journal of Consciousness Studies.*
www.imprint.co.uk/jcs.html

APPENDIX. The Fine Print

Physicalism and brain chemistry
"An earlier version of this was to suppose..." Penrose
did the heavy lifting here, I believe. He doesn't say
quantum uncertainty can be harnessed to provide freewill,
only that it precludes accurate prediction of the future,
hence we are not determined. Others have been less
scrupulous, I think, and seen uncertainty as a refuge for
God and soul. Roger Penrose, (1989). *The Emperor's New
Mind: Concerning Computers, Minds, and The Laws of
Physics.* Oxford University Press. Danah Zohar, (1990).

The Quantum Self: Human Nature and Consciousness Defined by the New Physics. Quill/William Morrow. Kenneth R. Miller, (1999). *Finding Darwin's God: A Scientist's Search for Common Ground Between God and Evolution.* Harper Perenial.

Evolutionary statistics

"In 1944 appeared an extremely influential book..." From a section titled "The Necessity of Mutation being a Rare Event," the full quote: "In order to be suitable material for the work of natural selection, mutations must be rare events, as they usually are... It is essential that they be introduced one at a time, while all the other parts of the mechanism are kept constant." Erwin Schrodinger, (1944). *What is Life?*

"The mathematicians were stunned..." Gordon Rattray Taylor, (1983). *The Great Evolution Mystery*, Harper & Row, p 4. Proceedings of the conference he quotes, "Mathematical Challenges to the Neo-Darwinian theory of Evolution," are available from The Wistar Institute.

"In a 1989 essay titled "The Limitations of ..." Maynard Smith, John. (1989). *Did Darwin Get It Right?* New York: Chapman and Hall.

"And in 2006 similar warnings dot a recent textbook ..." Massimo Pigliucci and Jonathan Kaplan (2006). *Making Sense of Evolution: The Conceptual Foundations of Evolutionary biology.* Chicago: The University of Chicago Press.

Disposing of more "epicycles"

"fossil records show that elephants..." *The Economist* (June 23 2007), page 92.

"...evolution is speeding along very nicely." For evolution told backwards from the present, marked in millions of years rather than in epochs, Richard Dawkins (2004). *The Ancestor's Tale: A Pilgrimage to the Dawn* *of Evolution.* Boston: Houghton Mifflin Company. Confirmed in George Gaylord Simpson (1949). *The Meaning of Evolution: A Study of the History of Live and Its Significance for Man.* Yale University Press": "...there is less complete but still sufficient evidence for the further generalization that the vertebrates have tended to evolve (structurally) faster than the invertebrates." P. 100. "If you look at the life of the Cambrian and then at that of today, the first and deepest impression is of increase....Even in the seas the increase is richly evident." P. 113.

"For another example think about whales..." Carl Zimmer, (1998). *At the Water's Edge: Fish with fingers, Whales with Legs, and How Life Came Ashore but Then Went Back to Sea.* Touchstone Book, Simon & Schuster.

"Here are extracts from a research paper." Tamames, J. (2001). Evolution of gene order conservation in prokaryotes, *Genome Biology*, 2.

"Wallace, co-discover of natural selection, turns against it" Andrew Berry (2002). *Infinite Tropics: An Alfred Wallace Anthology.* London: Verso.

"Comparing their world with the world he grew up in..." Peter Raby (2001). *Alfred Russel Wallace: A Life.* Princeton University Press.

Index

This index covers pages 1- 128

Me and The Genies

Cynical TV executive Henry Lazaard ("Lizard") is made manager of beautiful Sung-Tin Chi, Chinese scriptwriter for a children's cartoon series. Turns out, Sung-Tin's inspiration is evolution, specifically a mysterious document called "Beths Book." Follow their romance as Henry works to sabotage Sung-Tin's attempt to reform him. Along the way Sung-Tin reveals a plan for world domination that hangs on a clash between two concepts of the self, and explains evolution's role in forming them.

A quick-read introduction for teachers and school board members to the parties and issues involved in the controversy over teaching evolution in the classroom.

220 pages, 5 1/2 x 8 1/2. Paperback. USA $14.95.

ISBN 0-9779470-0-9. Available through Amazon.

Other titles published by

Evolved Self Publishing:

Me and The Genies

Father, in a Far Distant Time
I Find You

Father, in a Far Distant Time I Find You

It is 6991AD. Drawing on his father's wisdom about our time and our future, student Gregory Dumont shows us human nature being transformed by successive stages in evolutionary theory during a 2000-year-long "Age of World Figures" starting about 1000 years in our future. Combining a broad historical sensibility with respect for the discipline of scientific discourse, "Father..." reveals the implications for human nature of evolutionary theory over the past two centuries, readying us to assess the implications of steps to come. The take-home message—evolutionary theory shapes us as much as we shape it. Fine addition to the multi-disciplinary study of evolution.

202 pages, 5 1/2 x 8 1/2. Paperback. USA $21.95.
ISBN 0-9779470-1-7. Available through Amazon.

Father,
in a far
Distant Time
I Find You

Shaun Johnston

9 780977 947027